Rotorcraft Flying Handbook

Federal Aviation Administration

Skyhorse Publishing

www.skyhorsepublishing.com

10 9 8 7 6 5 4 3 2 1

ISBN-13: 978-1-60239-060-7
ISBN-10: 1-60239-060-6

Library of Congress Cataloging-in-Publication-Data is available on file.

Printed in China

PREFACE

The *Rotorcraft Flying Handbook* is designed as a technical manual for applicants who are preparing for their private, commercial, or flight instructor pilot certificates with a helicopter or gyroplane class rating. Certificated flight instructors may find this handbook a valuable training aid, since detailed coverage of aerodynamics, flight controls, systems, performance, flight maneuvers, emergencies, and aeronautical decision making is included. Topics, such as weather, navigation, radio navigation and communications, use of flight information publications, and regulations are available in other Federal Aviation Administration (FAA) publications.

This handbook conforms to pilot training and certification concepts established by the FAA. There are different ways of teaching, as well as performing flight procedures and maneuvers, and many variations in the explanations of aerodynamic theories and principles. This handbook adopts a selective method and concept to flying helicopters and gyroplanes. The discussion and explanations reflect the most commonly used practices and principles. Occasionally, the word "must" or similar language is used where the desired action is deemed critical. The use of such language is not intended to add to, interpret, or relieve a duty imposed by Title 14 of the Code of Federal Regulations (14 CFR). This handbook is divided into two parts. The first part, chapters 1 through 14, covers helicopters, and the second part, chapters 15 through 22, covers gyroplanes. The glossary and index apply to both parts.

It is essential for persons using this handbook to also become familiar with and apply the pertinent parts of 14 CFR and the *Aeronautical Information Manual (AIM)*. Performance standards for demonstrating competence required for pilot certification are prescribed in the appropriate rotorcraft practical test standard.

This handbook supersedes Advisory Circular (AC) 61-13B, *Basic Helicopter Handbook*, dated 1978. In addition, all or part of the information contained in the following advisory circulars are included in this handbook: AC 90-87, *Helicopter Dynamic Rollover;* AC 90-95, *Unanticipated Right Yaw in Helicopters;* AC 91-32B, *Safety in and around Helicopters;* and AC 91-42D, *Hazards of Rotating Propeller and Helicopter Rotor Blades.*

This publication may be purchased from the Superintendent of Documents, U.S. Government Printing Office (GPO), Washington, DC 20402-9325, or from U.S. Government Bookstores located in major cities throughout the United States.

The current Flight Standards Service airman training and testing material and subject matter knowledge codes for all airman certificates and ratings can be obtained from the Flight Standards Services web site at **http://av-info.faa.gov**.

Comments regarding this handbook should be sent to U.S. Department of Transportation, Federal Aviation Administration, Airman Testing Standards Branch, AFS-630, P.O. Box 25082, Oklahoma City, OK 73125.

AC 00-2, *Advisory Circular Checklist*, transmits the current status of FAA advisory circulars and other flight information publications. This checklist is free of charge and may be obtained by sending a request to U.S. Department of Transportation, Subsequent Distribution Office, SVC-121.23, Ardmore East Business Center, 3341 Q 75th Avenue, Landover, MD 20785.

AC00-2 also is available on the Internet at **http://www.faa.gov/abc/ac-chklst/actoc.htm**.

CONTENTS

HELICOPTER

GYROPLANE

Introduction to the Helicopter

Helicopters come in many sizes and shapes, but most share the same major components. These components include a cabin where the **payload** and crew are carried; an airframe, which houses the various components, or where components are attached; a powerplant or engine; and a transmission, which, among other things, takes the power from the engine and transmits it to the main rotor, which provides the aerodynamic forces that make the helicopter fly. Then, to keep the helicopter from turning due to **torque**, there must be some type of antitorque system. Finally there is the landing gear, which could be skids, wheels, skis, or floats. This chapter is an introduction to these components. [Figure 1-1]

THE MAIN ROTOR SYSTEM
The rotor system found on helicopters can consist of a single main rotor or dual rotors. With most dual rotors, the rotors turn in opposite directions so the torque from one rotor is opposed by the torque of the other. This cancels the turning tendencies. [Figure 1-2]

In general, a rotor system can be classified as either fully articulated, semirigid, or rigid. There are variations and combinations of these systems, which will be discussed in greater detail in Chapter 5—Helicopter Systems.

FULLY ARTICULATED ROTOR SYSTEM
A fully articulated rotor system usually consists of three or more rotor blades. The blades are allowed to **flap**, **feather**, and **lead or lag** independently of each other. Each rotor blade is attached to the rotor hub by a horizontal hinge, called the flapping hinge, which permits the blades to flap up and down. Each blade can move up and down independently of the others. The flapping hinge may be located at varying distances from the rotor hub, and there may be more than one. The position is chosen by each manufacturer, primarily with regard to stability and control.

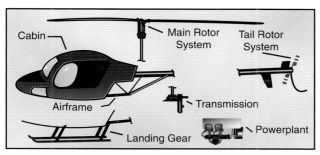

Figure 1-1. The major components of a helicopter are the cabin, airframe, landing gear, powerplant, transmission, main rotor system, and tail rotor system.

Figure 1-2. Helicopters can have a single main rotor or a dual rotor system.

Payload—The term used for passengers, baggage, and cargo.

Torque—In helicopters with a single, main rotor system, the tendency of the helicopter to turn in the opposite direction of the main rotor rotation.

Blade Flap—The upward or downward movement of the rotor blades during rotation.

Blade Feather or Feathering—The rotation of the blade around the spanwise (pitch change) axis.

Blade Lead or Lag—The fore and aft movement of the blade in the plane of rotation. It is sometimes called hunting or dragging.

Each rotor blade is also attached to the hub by a vertical hinge, called a drag or lag hinge, that permits each blade, independently of the others, to move back and forth in the plane of the rotor disc. Dampers are normally incorporated in the design of this type of rotor system to prevent excessive motion about the drag hinge. The purpose of the drag hinge and dampers is to absorb the acceleration and deceleration of the rotor blades.

The blades of a fully articulated rotor can also be feathered, or rotated about their spanwise axis. To put it more simply, feathering means the changing of the pitch angle of the rotor blades.

SEMIRIGID ROTOR SYSTEM

A semirigid rotor system allows for two different movements, flapping and feathering. This system is normally comprised of two blades, which are rigidly attached to the rotor hub. The hub is then attached to the rotor mast by a trunnion bearing or teetering hinge. This allows the blades to see-saw or flap together. As one blade flaps down, the other flaps up. Feathering is accomplished by the feathering hinge, which changes the pitch angle of the blade.

RIGID ROTOR SYSTEM

The rigid rotor system is mechanically simple, but structurally complex because operating loads must be absorbed in bending rather than through hinges. In this system, the blades cannot flap or lead and lag, but they can be feathered.

ANTITORQUE SYSTEMS
TAIL ROTOR

Most helicopters with a single, main rotor system require a separate rotor to overcome torque. This is accomplished through a variable pitch, antitorque rotor or tail rotor. [Figure 1-3]. You will need to vary the

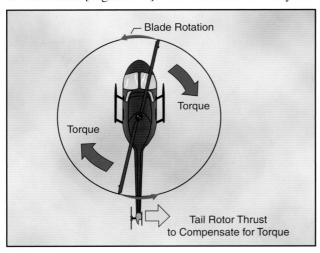

Figure 1-3. The antitorque rotor produces thrust to oppose torque and helps prevent the helicopter from turning in the opposite direction of the main rotor.

thrust of the antitorque system to maintain directional control whenever the main rotor torque changes, or to make heading changes while hovering.

FENESTRON

Another form of antitorque rotor is the fenestron or "fan-in-tail" design. This system uses a series of rotating blades shrouded within a vertical tail. Because the blades are located within a circular duct, they are less likely to come into contact with people or objects. [Figure 1-4]

Figure 1-4. Compared to an unprotected tail rotor, the fenestron antitorque system provides an improved margin of safety during ground operations.

NOTAR®

The NOTAR® system is an alternative to the antitorque rotor. The system uses low-pressure air that is forced into the tailboom by a fan mounted within the helicopter. The air is then fed through horizontal slots, located on the right side of the tailboom, and to a controllable rotating nozzle to provide antitorque and directional control. The low-pressure air coming from the horizontal slots, in conjunction with the downwash from the main rotor, creates a phenomenon called "Coanda Effect," which produces a lifting force on the right side of the tailboom. [Figure 1-5]

LANDING GEAR

The most common landing gear is a skid type gear, which is suitable for landing on various types of surfaces. Some types of skid gear are equipped with dampers so touchdown shocks or jolts are not transmitted to the main rotor system. Other types absorb the shocks by the bending of the skid attachment arms. Landing skids may be fitted with replaceable heavy-duty skid shoes to protect them from excessive wear and tear.

Helicopters can also be equipped with floats for water operations, or skis for landing on snow or soft terrain. Wheels are another type of landing gear. They may be in a tricycle or four point configuration. Normally, the

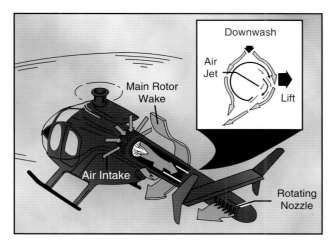

Figure 1-5. While in a hover, Coanda Effect supplies approximately two-thirds of the lift necessary to maintain directional control. The rest is created by directing the thrust from the controllable rotating nozzle.

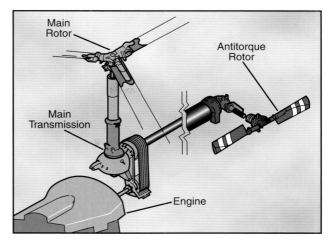

Figure 1-6. Typically, the engine drives the main rotor through a transmission and belt drive or centrifugal clutch system. The antitorque rotor is driven from the transmission.

nose or tail gear is free to swivel as the helicopter is taxied on the ground.

POWERPLANT

A typical small helicopter has a reciprocating engine, which is mounted on the airframe. The engine can be mounted horizontally or vertically with the transmission supplying the power to the vertical main rotor shaft. [Figure 1-6]

Another engine type is the gas turbine. This engine is used in most medium to heavy lift helicopters due to its large horsepower output. The engine drives the main transmission, which then transfers power directly to the main rotor system, as well as the tail rotor.

FLIGHT CONTROLS

When you begin flying a helicopter, you will use four basic flight controls. They are the cyclic pitch control; the collective pitch control; the throttle, which is usually a twist grip control located on the end of the collective lever; and the antitorque pedals. The collective and cyclic controls the pitch of the main rotor blades. The function of these controls will be explained in detail in Chapter 4—Flight Controls. [Figure 1-7]

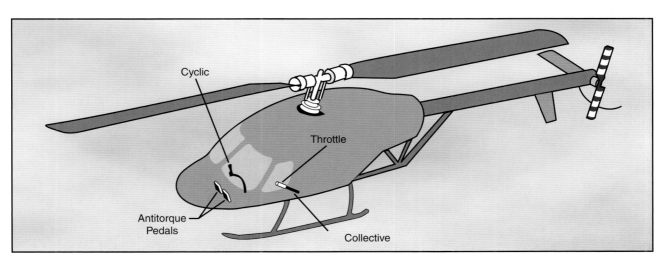

Figure 1-7. Location of flight controls.

There are four forces acting on a helicopter in flight. They are lift, weight, thrust, and drag. [Figure 2-1] Lift is the upward force created by the effect of airflow as it passes around an airfoil. Weight opposes lift and is caused by the downward pull of gravity. Thrust is the force that propels the helicopter through the air. Opposing lift and thrust is drag, which is the retarding force created by development of lift and the movement of an object through the air.

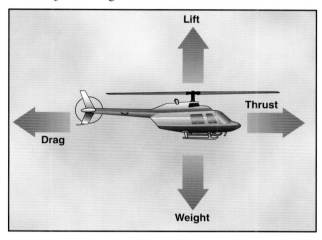

Figure 2-1. Four forces acting on a helicopter in forward flight.

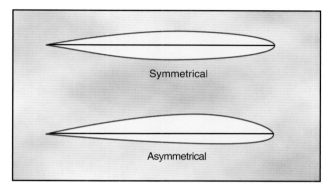

Figure 2-2. The upper and lower curvatures are the same on a symmetrical airfoil and vary on an asymmetrical airfoil.

AIRFOIL

Before beginning the discussion of lift, you need to be aware of certain aerodynamic terms that describe an airfoil and the interaction of the airflow around it.

An airfoil is any surface, such as an airplane wing or a helicopter rotor blade, which provides aerodynamic force when it interacts with a moving stream of air. Although there are many different rotor blade airfoil designs, in most helicopter flight conditions, all airfoils perform in the same manner.

Engineers of the first helicopters designed relatively thick airfoils for their structural characteristics. Because the rotor blades were very long and slender, it was necessary to incorporate more structural rigidity into them. This prevented excessive blade droop when the rotor system was idle, and minimized blade twisting while in flight. The airfoils were also designed to be symmetrical, which means they had the same camber (curvature) on both the upper and lower surfaces.

Symmetrical blades are very stable, which helps keep blade twisting and flight control loads to a minimum. [Figure 2-2] This stability is achieved by keeping the center of pressure virtually unchanged as the angle of attack changes. Center of pressure is the imaginary point on the chord line where the resultant of all aerodynamic forces are considered to be concentrated.

Today, designers use thinner airfoils and obtain the required rigidity by using composite materials. In addition, airfoils are asymmetrical in design, meaning the upper and lower surface do not have the same camber. Normally these airfoils would not be as stable, but this can be corrected by bending the trailing edge to produce the same characteristics as symmetrical airfoils. This is called "reflexing." Using this type of rotor blade allows the rotor system to operate at higher forward speeds.

One of the reasons an asymmetrical rotor blade is not as stable is that the center of pressure changes with changes in angle of attack. When the center of pressure lifting force is behind the pivot point on a rotor blade, it tends to cause the rotor disc to pitch up. As the angle of attack increases, the center of pressure moves forward. If it moves ahead of the pivot point, the pitch of the rotor disc decreases. Since the angle of attack of the rotor blades is constantly changing during each cycle of rotation, the blades tend to flap, feather, lead, and lag to a greater degree.

When referring to an airfoil, the span is the distance from the rotor hub to the blade tip. Blade twist refers to a changing chord line from the blade root to the tip.

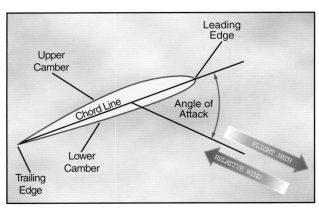

Figure 2-3. Aerodynamic terms of an airfoil.

Twisting a rotor blade causes it to produce a more even amount of lift along its span. This is necessary because rotational velocity increases toward the blade tip. The leading edge is the first part of the airfoil to meet the oncoming air. [Figure 2-3] The trailing edge is the aft portion where the airflow over the upper surface joins the airflow under the lower surface. The chord line is an imaginary straight line drawn from the leading to the trailing edge. The camber is the curvature of the airfoil's upper and lower surfaces. The relative wind is the wind moving past the airfoil. The direction of this wind is relative to the attitude, or position, of the airfoil and is always parallel, equal, and opposite in direction to the flight path of the airfoil. The angle of attack is the angle between the blade chord line and the direction of the relative wind.

RELATIVE WIND

Relative wind is created by the motion of an airfoil through the air, by the motion of air past an airfoil, or by a combination of the two. Relative wind may be affected by several factors, including the rotation of the rotor blades, horizontal movement of the helicopter, flapping of the rotor blades, and wind speed and direction.

For a helicopter, the relative wind is the flow of air with respect to the rotor blades. If the rotor is stopped, wind blowing over the blades creates a relative wind. When the helicopter is hovering in a no-wind condition, relative wind is created by the motion of the rotor blades through the air. If the helicopter is hovering in a wind, the relative wind is a combination of the wind and the motion of the rotor blades through the air. When the helicopter is in forward flight, the relative wind is a combination of the rotation of the rotor blades and the forward speed of the helicopter.

BLADE PITCH ANGLE

The pitch angle of a rotor blade is the angle between its chord line and the reference plane containing the rotor hub. [Figure 2-4] You control the pitch angle of the blades with the flight controls. The collective pitch changes each rotor blade an equal amount of pitch no matter where it is located in the plane of rotation (rotor disc) and is used to change rotor thrust. The cyclic pitch control changes the pitch of each blade as a function of where it is in the plane of rotation. This allows for trimming the helicopter in **pitch** and **roll** during forward flight and for maneuvering in all flight conditions.

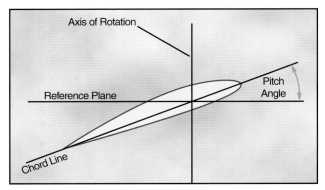

Figure 2-4. Do not confuse the axis of rotation with the rotor mast. The only time they coincide is when the tip-path plane is perpendicular to the rotor mast.

ANGLE OF ATTACK

When the angle of attack is increased, air flowing over the airfoil is diverted over a greater distance, resulting in an increase of air velocity and more lift. As angle of attack is increased further, it becomes more difficult for air to flow smoothly across the top of the airfoil. At this point the airflow begins to separate from the airfoil and enters a burbling or turbulent pattern. The turbulence results in a large increase in drag and loss of lift in the area where it is taking place. Increasing the angle of attack increases lift until the critical angle of attack is reached. Any increase in the angle of attack beyond this point produces a stall and a rapid decrease in lift. [Figure 2-5]

Angle of attack should not be confused with pitch angle. Pitch angle is determined by the direction of the relative wind. You can, however, change the angle of attack by changing the pitch angle through the use of the flight controls. If the pitch angle is increased, the angle of attack is increased, if the pitch angle is reduced, the angle of attack is reduced. [Figure 2-6]

Axis-of-Rotation—The imaginary line about which the rotor rotates. It is represented by a line drawn through the center of, and perpendicular to, the tip-path plane.

Tip-Path Plane—The imaginary circular plane outlined by the rotor blade tips as they make a cycle of rotation.

Aircraft Pitch—When referenced to a helicopter, is the movement of the helicopter about its lateral, or side to side axis. Movement of the cyclic forward or aft causes the nose of the helicopter to move up or down.

Aircraft Roll—Is the movement of the helicopter about its longitudinal, or nose to tail axis. Movement of the cyclic right or left causes the helicopter to tilt in that direction.

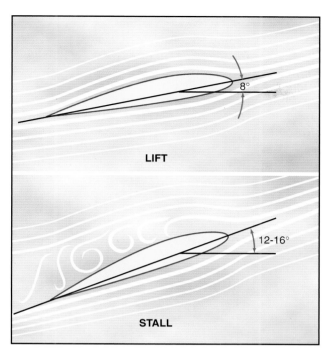

Figure 2-5. As the angle of attack is increased, the separation point starts near the trailing edge of the airfoil and progresses forward. Finally, the airfoil loses its lift and a stall condition occurs.

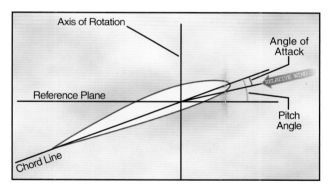

Figure 2-6. Angle of attack may be greater than, less than, or the same as the pitch angle.

LIFT
MAGNUS EFFECT

The explanation of lift can best be explained by looking at a cylinder rotating in an airstream. The local velocity near the cylinder is composed of the airstream velocity and the cylinder's rotational velocity, which decreases with distance from the cylinder. On a cylinder, which is rotating in such a way that the top surface area is rotating in the same direction as the airflow, the local velocity at the surface is high on top and low on the bottom.

As shown in figure 2-7, at point "A," a stagnation point exists where the airstream line that impinges on the surface splits; some air goes over and some under. Another stagnation point exists at "B," where the two air streams rejoin and resume at identical velocities. We

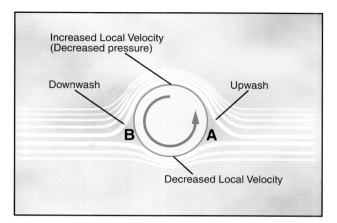

Figure 2-7. Magnus Effect is a lifting force produced when a rotating cylinder produces a pressure differential. This is the same effect that makes a baseball curve or a golf ball slice.

now have upwash ahead of the rotating cylinder and downwash at the rear.

The difference in surface velocity accounts for a difference in pressure, with the pressure being lower on the top than the bottom. This low pressure area produces an upward force known as the "Magnus Effect." This mechanically induced circulation illustrates the relationship between circulation and lift.

An airfoil with a positive angle of attack develops air circulation as its sharp trailing edge forces the rear stagnation point to be aft of the trailing edge, while the front stagnation point is below the leading edge. [Figure 2-8]

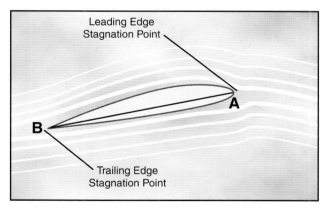

Figure 2-8. Air circulation around an airfoil occurs when the front stagnation point is below the leading edge and the aft stagnation point is beyond the trailing edge.

BERNOULLI'S PRINCIPLE

Air flowing over the top surface accelerates. The airfoil is now subjected to Bernoulli's Principle or the "venturi effect." As air velocity increases through the constricted portion of a venturi tube, the pressure decreases.

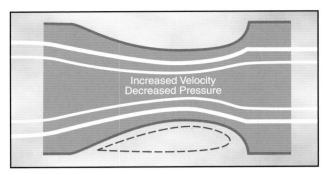

Figure 2-9. The upper surface of an airfoil is similar to the constriction in a venturi tube.

Compare the upper surface of an airfoil with the constriction in a venturi tube that is narrower in the middle than at the ends. [Figure 2-9]

The upper half of the venturi tube can be replaced by layers of undisturbed air. Thus, as air flows over the upper surface of an airfoil, the camber of the airfoil causes an increase in the speed of the airflow. The increased speed of airflow results in a decrease in pressure on the upper surface of the airfoil. At the same time, air flows along the lower surface of the airfoil, building up pressure. The combination of decreased pressure on the upper surface and increased pressure on the lower surface results in an upward force. [Figure 2-10]

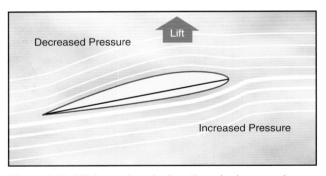

Figure 2-10. Lift is produced when there is decreased pressure above and increased pressure below an airfoil.

As angle of attack is increased, the production of lift is increased. More upwash is created ahead of the airfoil as the leading edge stagnation point moves under the leading edge, and more downwash is created aft of the trailing edge. Total lift now being produced is perpendicular to relative wind. In summary, the production of lift is based upon the airfoil creating circulation in the airstream (Magnus Effect) and creating differential pressure on the airfoil (Bernoulli's Principle).

NEWTON'S THIRD LAW OF MOTION
Additional lift is provided by the rotor blade's lower surface as air striking the underside is deflected down-

ward. According to Newton's Third Law of Motion, "for every action there is an equal and opposite reaction," the air that is deflected downward also produces an upward (lifting) reaction.

Since air is much like water, the explanation for this source of lift may be compared to the planing effect of skis on water. The lift which supports the water skis (and the skier) is the force caused by the impact pressure and the deflection of water from the lower surfaces of the skis.

Under most flying conditions, the impact pressure and the deflection of air from the lower surface of the rotor blade provides a comparatively small percentage of the total lift. The majority of lift is the result of decreased pressure above the blade, rather than the increased pressure below it.

WEIGHT
Normally, weight is thought of as being a known, fixed value, such as the weight of the helicopter, fuel, and occupants. To lift the helicopter off the ground vertically, the rotor system must generate enough lift to overcome or offset the total weight of the helicopter and its occupants. This is accomplished by increasing the pitch angle of the main rotor blades.

The weight of the helicopter can also be influenced by aerodynamic loads. When you bank a helicopter while maintaining a constant altitude, the "G" load or load factor increases. Load factor is the ratio of the load supported by the main rotor system to the actual weight of the helicopter and its contents. In **steady-state flight**, the helicopter has a load factor of one, which means the main rotor system is supporting the actual total weight of the helicopter. If you increase the bank angle to 60°, while still maintaining a constant altitude, the load factor increases to two. In this case, the main rotor system has to support twice the weight of the helicopter and its contents. [Figure 2-11]

Disc loading of a helicopter is the ratio of weight to the total main rotor disc area, and is determined by dividing the total helicopter weight by the rotor disc area, which is the area swept by the blades of a rotor. Disc area can be found by using the span of one rotor blade as the radius of a circle and then determining the area the blades encompass during a complete rotation. As the helicopter is maneuvered, disc loading changes. The higher the loading, the more power you need to maintain rotor speed.

Steady-State Flight—A condition when an aircraft is in straight-and-level, unaccelerated flight, and all forces are in balance.

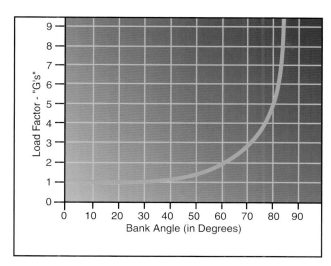

Figure 2-11. The load factor diagram allows you to calculate the amount of "G" loading exerted with various angle of bank.

THRUST

Thrust, like lift, is generated by the rotation of the main rotor system. In a helicopter, thrust can be forward, rearward, sideward, or vertical. The resultant of lift and thrust determines the direction of movement of the helicopter.

The solidity ratio is the ratio of the total rotor blade area, which is the combined area of all the main rotor blades, to the total rotor disc area. This ratio provides a means to measure the potential for a rotor system to provide thrust.

The tail rotor also produces thrust. The amount of thrust is variable through the use of the antitorque pedals and is used to control the helicopter's **yaw**.

DRAG

The force that resists the movement of a helicopter through the air and is produced when lift is developed is called drag. Drag always acts parallel to the relative wind. Total drag is composed of three types of drag: profile, induced, and parasite.

PROFILE DRAG

Profile drag develops from the frictional resistance of the blades passing through the air. It does not change significantly with the airfoil's angle of attack, but increases moderately when airspeed increases. Profile drag is composed of form drag and skin friction.

Form drag results from the turbulent wake caused by the separation of airflow from the surface of a structure. The amount of drag is related to both the size and shape of the structure that protrudes into the relative wind. [Figure 2-12]

Skin friction is caused by surface roughness. Even though the surface appears smooth, it may be quite rough when viewed under a microscope. A thin layer of air clings to the rough surface and creates small eddies that contribute to drag.

INDUCED DRAG

Induced drag is generated by the airflow circulation around the rotor blade as it creates lift. The high-pressure area beneath the blade joins the low-pressure air above the blade at the trailing edge and at the rotor tips. This causes a spiral, or vortex, which trails behind each blade whenever lift is being produced. These vortices deflect the airstream downward in the vicinity of the blade, creating an increase in downwash. Therefore, the blade operates in an average relative wind that is inclined downward and rearward near the blade. Because the lift produced by the blade is perpendicular

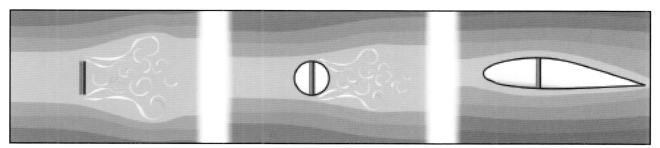

Figure 2-12. It is easy to visualize the creation of form drag by examining the airflow around a flat plate. Streamlining decreases form drag by reducing the airflow separation.

Aircraft Yaw—The movement of the helicopter about its vertical axis.

to the relative wind, the lift is inclined aft by the same amount. The component of lift that is acting in a rearward direction is induced drag. [Figure 2-13]

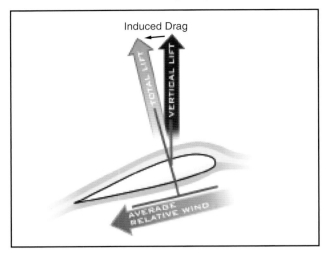

Figure 2-13. The formation of induced drag is associated with the downward deflection of the airstream near the rotor blade.

As the air pressure differential increases with an increase in angle of attack, stronger vortices form, and induced drag increases. Since the blade's angle of attack is usually lower at higher airspeeds, and higher at low speeds, induced drag decreases as airspeed increases and increases as airspeed decreases. Induced drag is the major cause of drag at lower airspeeds.

PARASITE DRAG
Parasite drag is present any time the helicopter is moving through the air. This type of drag increases with airspeed. Nonlifting components of the helicopter, such as the cabin, rotor mast, tail, and landing gear, contribute to parasite drag. Any loss of momentum by the airstream, due to such things as openings for engine cooling, creates additional parasite drag. Because of its rapid increase

with increasing airspeed, parasite drag is the major cause of drag at higher airspeeds. Parasite drag varies with the square of the velocity. Doubling the airspeed increases the parasite drag four times.

TOTAL DRAG
Total drag for a helicopter is the sum of all three drag forces. [Figure 2-14] As airspeed increases, parasite drag increases, while induced drag decreases. Profile drag remains relatively constant throughout the speed range with some increase at higher airspeeds. Combining all drag forces results in a total drag curve. The low point on the total drag curve shows the airspeed at which drag is minimized. This is the point where the lift-to-drag ratio is greatest and is referred to as L/D_{max}. At this speed, the total lift capacity of the helicopter, when compared to the total drag of the helicopter, is most favorable. This is important in helicopter performance.

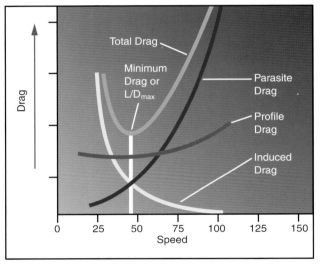

Figure 2-14. The total drag curve represents the combined forces of parasite, profile, and induced drag; and is plotted against airspeed.

L/D_{max}—The maximum ratio between total lift (L) and the total drag (D). This point provides the best glide speed. Any deviation from best glide speed increases drag and reduces the distance you can glide.

Aerodynamics of Flight

Once a helicopter leaves the ground, it is acted upon by the four aerodynamic forces. In this chapter, we will examine these forces as they relate to flight maneuvers.

POWERED FLIGHT

In powered flight (hovering, vertical, forward, sideward, or rearward), the total lift and thrust forces of a rotor are perpendicular to the tip-path plane or plane of rotation of the rotor.

HOVERING FLIGHT

For standardization purposes, this discussion assumes a stationary hover in a no-wind condition. During hovering flight, a helicopter maintains a constant position over a selected point, usually a few feet above the ground. For a helicopter to hover, the lift and thrust produced by the rotor system act straight up and must equal the weight and drag, which act straight down. While hovering, you can change the amount of main rotor thrust to maintain the desired hovering altitude. This is done by changing the angle of attack of the main rotor blades and by varying power, as needed. In this case, thrust acts in the same vertical direction as lift. [Figure 3-1]

The weight that must be supported is the total weight of the helicopter and its occupants. If the amount of thrust is greater than the actual weight, the helicopter gains altitude; if thrust is less than weight, the helicopter loses altitude.

The drag of a hovering helicopter is mainly induced drag incurred while the blades are producing lift. There is, however, some profile drag on the blades as they rotate through the air. Throughout the rest of this discussion, the term "drag" includes both induced and profile drag.

An important consequence of producing thrust is torque. As stated before, for every action there is an equal and opposite reaction. Therefore, as the engine turns the main rotor system in a counterclockwise direction, the helicopter fuselage turns clockwise. The amount of torque is directly related to the amount of engine power being used to turn the main rotor system. Remember, as power changes, torque changes.

To counteract this torque-induced turning tendency, an antitorque rotor or tail rotor is incorporated into most helicopter designs. You can vary the amount of thrust produced by the tail rotor in relation to the amount of torque produced by the engine. As the engine supplies more power, the tail rotor must produce more thrust. This is done through the use of antitorque pedals.

TRANSLATING TENDENCY OR DRIFT

During hovering flight, a single main rotor helicopter tends to drift in the same direction as antitorque rotor thrust. This drifting tendency is called translating tendency. [Figure 3-2]

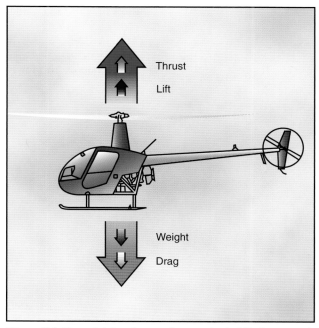

Figure 3-1. To maintain a hover at a constant altitude, enough lift and thrust must be generated to equal the weight of the helicopter and the drag produced by the rotor blades.

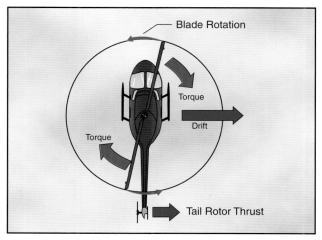

Figure 3-2. A tail rotor is designed to produce thrust in a direction opposite torque. The thrust produced by the tail rotor is sufficient to move the helicopter laterally.

To counteract this drift, one or more of the following features may be used:

- The main transmission is mounted so that the rotor mast is rigged for the tip-path plane to have a built-in tilt opposite tail thrust, thus producing a small sideward thrust.

- Flight control rigging is designed so that the rotor disc is tilted slightly opposite tail rotor thrust when the cyclic is centered.

- The cyclic pitch control system is designed so that the rotor disc tilts slightly opposite tail rotor thrust when in a hover.

Counteracting translating tendency, in a helicopter with a counterclockwise main rotor system, causes the left skid to hang lower while hovering. The opposite is true for rotor systems turning clockwise when viewed from above.

PENDULAR ACTION

Since the fuselage of the helicopter, with a single main rotor, is suspended from a single point and has considerable mass, it is free to oscillate either longitudinally or laterally in the same way as a pendulum. This pendular action can be exaggerated by over controlling; therefore, control movements should be smooth and not exaggerated. [Figure 3-3]

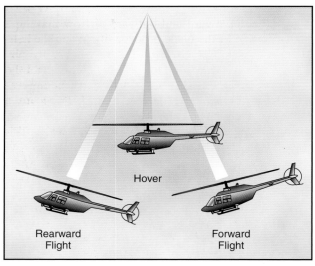

Figure 3-3. Because the helicopter's body has mass and is suspended from a single point (the rotor mast head), it tends to act much like a pendulum.

CONING

In order for a helicopter to generate lift, the rotor blades must be turning. This creates a relative wind that is opposite the direction of rotor system rotation. The rotation of the rotor system creates **centrifugal force** (inertia), which tends to pull the blades straight outward from the main rotor hub. The faster the rotation, the

greater the centrifugal force. This force gives the rotor blades their rigidity and, in turn, the strength to support the weight of the helicopter. The centrifugal force generated determines the maximum operating rotor r.p.m. due to structural limitations on the main rotor system.

As a vertical takeoff is made, two major forces are acting at the same time—centrifugal force acting outward and perpendicular to the rotor mast, and lift acting upward and parallel to the mast. The result of these two forces is that the blades assume a conical path instead of remaining in the plane perpendicular to the mast. [Figure 3-4]

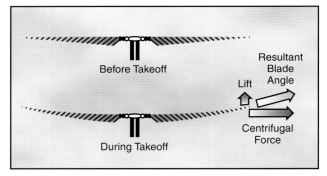

Figure 3-4. Rotor blade coning occurs as the rotor blades begin to lift the weight of the helicopter. In a semirigid and rigid rotor system, coning results in blade bending. In an articulated rotor system, the blades assume an upward angle through movement about the flapping hinges.

CORIOLIS EFFECT (LAW OF CONSERVATION OF ANGULAR MOMENTUM)

Coriolis Effect, which is sometimes referred to as conservation of angular momentum, might be compared to spinning skaters. When they extend their arms, their rotation slows down because the center of mass moves farther from the axis of rotation. When their arms are retracted, the rotation speeds up because the center of mass moves closer to the axis of rotation.

When a rotor blade flaps upward, the center of mass of that blade moves closer to the axis of rotation and blade acceleration takes place in order to conserve angular momentum. Conversely, when that blade flaps downward, its center of mass moves further from the axis of

Centrifugal Force—The apparent force that an object moving along a circular path exerts on the body constraining the obect and that acts outwardly away from the center of rotation.

rotation and blade deceleration takes place. [Figure 3-5] Keep in mind that due to coning, a rotor blade will not flap below a plane passing through the rotor hub and perpendicular to the axis of rotation. The acceleration and deceleration actions of the rotor blades are absorbed by either dampers or the blade structure itself, depending upon the design of the rotor system.

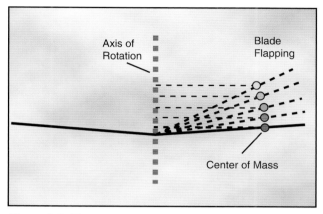

Figure 3-5. The tendency of a rotor blade to increase or decrease its velocity in its plane of rotation due to mass movement is known as Coriolis Effect, named for the mathematician who made studies of forces generated by radial movements of mass on a rotating disc.

Two-bladed rotor systems are normally subject to Coriolis Effect to a much lesser degree than are articulated rotor systems since the blades are generally "underslung" with respect to the rotor hub, and the change in the distance of the center of mass from the axis of rotation is small. [Figure 3-6] The hunting action is absorbed by the blades through bending. If a two-bladed rotor system is not "underslung," it will be

subject to Coriolis Effect comparable to that of a fully articulated system.

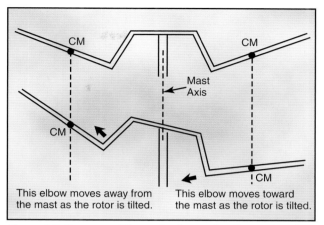

Figure 3-6. Because of the underslung rotor, the center of mass remains approximately the same distance from the mast after the rotor is tilted.

GROUND EFFECT

When hovering near the ground, a phenomenon known as ground effect takes place. [Figure 3-7] This effect usually occurs less than one rotor diameter above the surface. As the induced airflow through the rotor disc is reduced by the surface friction, the lift vector increases. This allows a lower rotor blade angle for the same amount of lift, which reduces induced drag. Ground effect also restricts the generation of blade tip vortices due to the downward and outward airflow making a larger portion of the blade produce lift. When the helicopter gains altitude vertically, with no forward airspeed, induced airflow is no longer restricted, and the blade tip vortices increase with the decrease in outward airflow. As a result, drag increases which means a

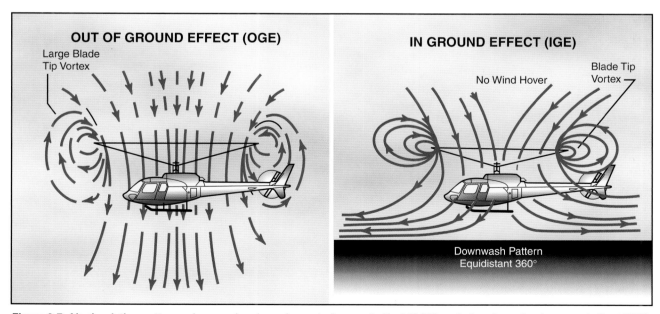

Figure 3-7. Air circulation patterns change when hovering out of ground effect (OGE) and when hovering in ground effect (IGE).

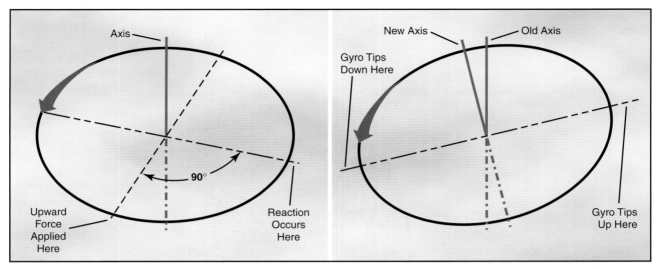

Figure 3-8. Gyroscopic precession principle—when a force is applied to a spinning gyro, the maximum reaction occurs approximately 90° later in the direction of rotation.

higher pitch angle, and more power is needed to move the air down through the rotor.

Ground effect is at its maximum in a no-wind condition over a firm, smooth surface. Tall grass, rough terrain, revetments, and water surfaces alter the airflow pattern, causing an increase in rotor tip vortices.

GYROSCOPIC PRECESSION
The spinning main rotor of a helicopter acts like a gyroscope. As such, it has the properties of gyroscopic action, one of which is precession. Gyroscopic precession is the resultant action or deflection of a spinning object when a force is applied to this object. This action occurs approximately 90° in the direction of rotation from the point where the force is applied. [Figure 3-8]

Let us look at a two-bladed rotor system to see how gyroscopic precession affects the movement of the tip-path plane. Moving the cyclic pitch control increases the angle of attack of one rotor blade with the result that a greater lifting force is applied at that point in the plane of rotation. This same control movement simultaneously decreases the angle of attack of the other blade the same amount, thus decreasing the lifting force applied at that point in the plane of rotation. The blade with the increased angle of attack tends to flap up; the blade with the decreased angle of attack tends to flap down. Because the rotor disk acts like a gyro, the blades reach maximum deflection at a point approximately 90° later in the plane of rotation. As shown in figure 3-9, the retreating blade angle of attack is increased and the advancing blade angle of attack is decreased resulting in a tipping forward of the tip-path plane, since maximum deflection takes place 90° later when the blades are at the rear and front, respectively.

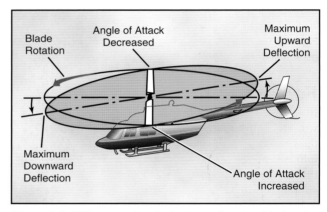

Figure 3-9. With a counterclockwise main rotor blade rotation, as each blade passes the 90° position on the left, the maximum increase in angle of attack occurs. As each blade passes the 90° position to the right, the maximum decrease in angle of attack occurs. Maximum deflection takes place 90° later—maximum upward deflection at the rear and maximum downward deflection at the front—and the tip-path plane tips forward.

In a rotor system using three or more blades, the movement of the cyclic pitch control changes the angle of attack of each blade an appropriate amount so that the end result is the same.

VERTICAL FLIGHT
Hovering is actually an element of vertical flight. Increasing the angle of attack of the rotor blades (pitch) while their velocity remains constant generates additional vertical lift and thrust and the helicopter ascends. Decreasing the pitch causes the helicopter to descend. In a no wind condition when lift and thrust are less than weight and drag, the helicopter descends vertically. If

lift and thrust are greater than weight and drag, the helicopter ascends vertically. [Figure 3-10]

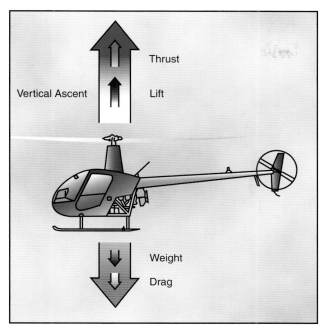

Figure 3-10. To ascend vertically, more lift and thrust must be generated to overcome the forces of weight and the drag.

FORWARD FLIGHT

In or during forward flight, the tip-path plane is tilted forward, thus tilting the total lift-thrust force forward from the vertical. This resultant lift-thrust force can be resolved into two components—lift acting vertically upward and thrust acting horizontally in the direction of flight. In addition to lift and thrust, there is weight (the downward acting force) and drag (the rearward acting or retarding force of inertia and wind resistance). [Figure 3-11]

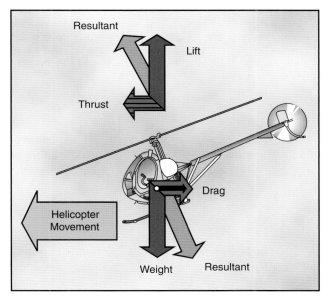

Figure 3-11. To transition into forward flight, some of the vertical thrust must be vectored horizontally. You initiate this by forward movement of the cyclic control.

In straight-and-level, unaccelerated forward flight, lift equals weight and thrust equals drag (straight-and-level flight is flight with a constant heading and at a constant altitude). If lift exceeds weight, the helicopter climbs; if lift is less than weight, the helicopter descends. If thrust exceeds drag, the helicopter speeds up; if thrust is less than drag, it slows down.

As the helicopter moves forward, it begins to lose altitude because of the lift that is lost as thrust is diverted forward. However, as the helicopter begins to accelerate, the rotor system becomes more efficient due to the increased airflow. The result is excess power over that which is required to hover. Continued acceleration causes an even larger increase in airflow through the rotor disc and more excess power.

TRANSLATIONAL LIFT

Translational lift is present with any horizontal flow of air across the rotor. This increased flow is most noticeable when the airspeed reaches approximately 16 to 24 knots. As the helicopter accelerates through this speed, the rotor moves out of its vortices and is in relatively undisturbed air. The airflow is also now more horizontal, which reduces induced flow and drag with a corresponding increase in angle of attack and lift. The additional lift available at this speed is referred to as "effective translational lift" (ETL). [Figure 3-12]

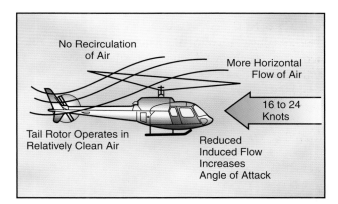

Figure 3-12. Effective translational lift is easily recognized in actual flight by a transient induced aerodynamic vibration and increased performance of the helicopter.

When a single-rotor helicopter flies through translational lift, the air flowing through the main rotor and over the tail rotor becomes less turbulent and more aerodynamically efficient. As the tail rotor efficiency improves, more thrust is produced causing the aircraft to yaw left in a counterclockwise rotor system. It will be necessary to use right torque pedal to correct for this tendency on takeoff. Also, if no corrections are made, the nose rises or pitches up, and rolls to the right. This is caused by combined effects of dissymmetry of lift and transverse flow effect, and is corrected with cyclic control.

Translational lift is also present in a stationary hover if the wind speed is approximately 16 to 24 knots. In normal operations, always utilize the benefit of translational lift, especially if maximum performance is needed.

INDUCED FLOW

As the rotor blades rotate they generate what is called rotational relative wind. This airflow is characterized as flowing parallel and opposite the rotor's plane of rotation and striking perpendicular to the rotor blade's leading edge. This rotational relative wind is used to generate lift. As rotor blades produce lift, air is accelerated over the foil and projected downward. Anytime a helicopter is producing lift, it moves large masses of air vertically and down through the rotor system. This downwash or induced flow can significantly change the efficiency of the rotor system. Rotational relative wind combines with induced flow to form the resultant relative wind. As induced flow increases, resultant relative wind becomes less horizontal. Since angle of attack is determined by measuring the difference between the chord line and the resultant relative wind, as the resultant relative wind becomes less horizontal, angle of attack decreases. [Figure 3-13]

TRANSVERSE FLOW EFFECT

As the helicopter accelerates in forward flight, induced flow drops to near zero at the forward disc area and increases at the aft disc area. This increases the angle of attack at the front disc area causing the rotor blade to flap up, and reduces angle of attack at the aft disc area causing the rotor blade to flap down. Because the rotor acts like a gyro, maximum displacement occurs 90° in the direction of rotation. The result is a tendency for the helicopter to roll slightly to the right as it accelerates through approximately 20 knots or if the headwind is approximately 20 knots.

You can recognize transverse flow effect because of increased vibrations of the helicopter at airspeeds just below effective translational lift on takeoff and after passing through effective translational lift during landing. To counteract transverse flow effect, a cyclic input needs to be made.

DISSYMMETRY OF LIFT

When the helicopter moves through the air, the relative airflow through the main rotor disc is different on the advancing side than on the retreating side. The relative wind encountered by the advancing blade is increased by the forward speed of the helicopter, while the relative wind speed acting on the retreating blade is reduced by the helicopter's forward airspeed. Therefore, as a result of the relative wind speed, the advancing blade side of the rotor disc produces more lift than the retreating blade side. This situation is defined as dissymmetry of lift. [Figure 3-14]

If this condition was allowed to exist, a helicopter with a counterclockwise main rotor blade rotation would roll to the left because of the difference in lift. In reality, the main rotor blades flap and feather automatically to equalize lift across the rotor disc. Articulated rotor systems, usually with three or more blades, incorporate a horizontal hinge (flapping hinge) to allow the individual rotor blades to move, or flap up and down as they rotate. A semirigid rotor system (two blades) utilizes a teetering hinge, which allows the blades to flap as a unit. When one blade flaps up, the other flaps down.

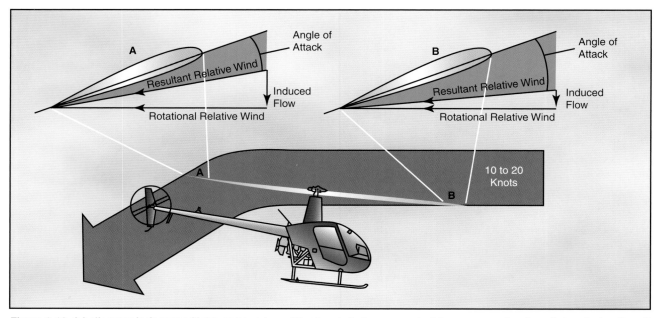

Figure 3-13. A helicopter in forward flight, or hovering with a headwind or crosswind, has more molecules of air entering the aft portion of the rotor blade. Therefore, the angle of attack is less and the induced flow is greater at the rear of the rotor disc.

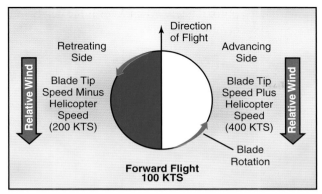

Figure 3-14. The blade tip speed of this helicopter is approximately 300 knots. If the helicopter is moving forward at 100 knots, the relative wind speed on the advancing side is 400 knots. On the retreating side, it is only 200 knots. This difference in speed causes a dissymmetry of lift.

As shown in figure 3-15, as the rotor blade reaches the advancing side of the rotor disc (A), it reaches its maximum upflap velocity. When the blade flaps upward, the angle between the chord line and the resultant relative wind decreases. This decreases the angle of attack, which reduces the amount of lift produced by the blade. At position (C) the rotor blade is now at its maximum downflapping velocity. Due to downflapping, the angle between the chord line and the resultant relative wind increases. This increases the angle of attack and thus the amount of lift produced by the blade.

The combination of blade flapping and slow relative wind acting on the retreating blade normally limits the maximum forward speed of a helicopter. At a high forward speed, the retreating blade stalls because of a high angle of attack and slow relative wind speed. This situation is called retreating blade stall and is evidenced by a nose pitch up, vibration, and a rolling tendency—usually to the left in helicopters with counterclockwise blade rotation.

You can avoid retreating blade stall by not exceeding the never-exceed speed. This speed is designated V_{NE} and is usually indicated on a placard and marked on the airspeed indicator by a red line.

During aerodynamic flapping of the rotor blades as they compensate for dissymmetry of lift, the advancing blade

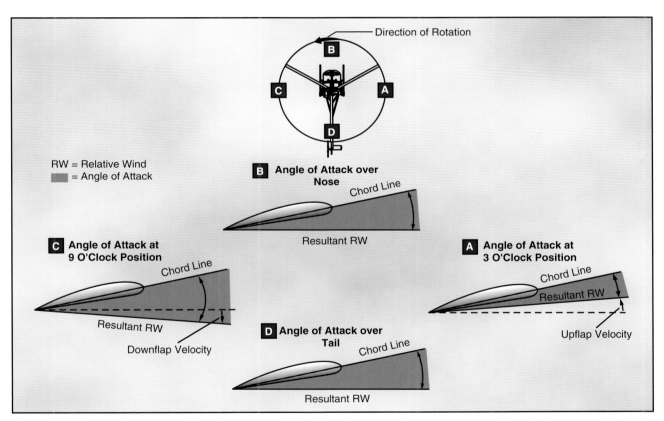

Figure 3-15. The combined upward flapping (reduced lift) of the advancing blade and downward flapping (increased lift) of the retreating blade equalizes lift across the main rotor disc counteracting dissymmetry of lift.

V_{NE}—The speed beyond which an aircraft should never be operated. V_{NE} can change with altitude, density altitude, and weight.

achieves maximum upflapping displacement over the nose and maximum downflapping displacement over the tail. This causes the tip-path plane to tilt to the rear and is referred to as blowback. Figure 3-16 shows how the rotor disc was originally oriented with the front down following the initial cyclic input, but as airspeed is gained and flapping eliminates dissymmetry of lift, the front of the disc comes up, and the back of the disc goes down. This reorientation of the rotor disc changes the direction in which total rotor thrust acts so that the helicopter's forward speed slows, but can be corrected with cyclic input.

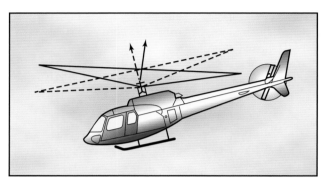

Figure 3-16. To compensate for blowback, you must move the cyclic forward. Blowback is more pronounced with higher airspeeds.

SIDEWARD FLIGHT

In sideward flight, the tip-path plane is tilted in the direction that flight is desired. This tilts the total lift-thrust vector sideward. In this case, the vertical or lift component is still straight up and weight straight down, but the horizontal or thrust component now acts sideward with drag acting to the opposite side. [Figure 3-17]

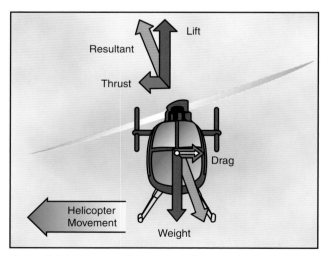

Figure 3-17. Forces acting on the helicopter during sideward flight.

REARWARD FLIGHT

For rearward flight, the tip-path plane is tilted rearward, which, in turn, tilts the lift-thrust vector rear-

ward. Drag now acts forward with the lift component straight up and weight straight down. [Figure 3-18]

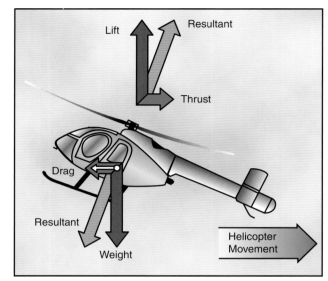

Figure 3-18. Forces acting on the helicopter during rearward flight.

TURNING FLIGHT

In forward flight, the rotor disc is tilted forward, which also tilts the total lift-thrust force of the rotor disc forward. When the helicopter is banked, the rotor disc is tilted sideward resulting in lift being separated into two components. Lift acting upward and opposing weight is called the vertical component of lift. Lift acting horizontally and opposing inertia (centrifugal force) is the horizontal component of lift (**centripetal force**). [Figure 3-19]

As the angle of bank increases, the total lift force is tilted more toward the horizontal, thus causing the rate of turn to increase because more lift is acting horizontally. Since the resultant lifting force acts more horizontally, the effect of lift acting vertically is deceased. To compensate for this decreased vertical lift, the angle of attack of the rotor blades must be increased in order to maintain altitude. The steeper the angle of bank, the greater the angle of attack of the rotor blades required to maintain altitude. Thus, with an increase in bank and a greater angle of attack, the resultant lifting force increases and the rate of turn is faster.

AUTOROTATION

Autorotation is the state of flight where the main rotor system is being turned by the action of relative wind

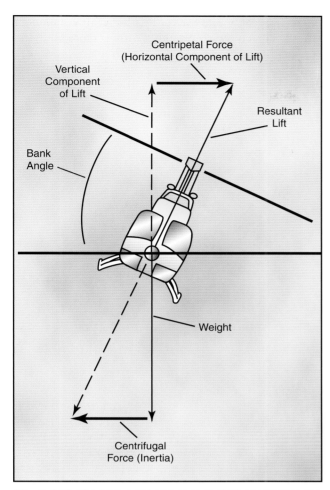

Figure 3-19. The horizontal component of lift accelerates the helicopter toward the center of the turn.

rather than engine power. It is the means by which a helicopter can be landed safely in the event of an engine failure. In this case, you are using altitude as potential energy and converting it to kinetic energy during the descent and touchdown. All helicopters must have this capability in order to be certified. Autorotation is permitted mechanically because of a freewheeling unit, which allows the main rotor to con-

tinue turning even if the engine is not running. In normal powered flight, air is drawn into the main rotor system from above and exhausted downward. During autorotation, airflow enters the rotor disc from below as the helicopter descends. [Figure 3-20]

AUTOROTATION (VERTICAL FLIGHT)

Most autorotations are performed with forward speed. For simplicity, the following aerodynamic explanation is based on a vertical autorotative descent (no forward speed) in still air. Under these conditions, the forces that cause the blades to turn are similar for all blades regardless of their position in the plane of rotation. Therefore, dissymmetry of lift resulting from helicopter airspeed is not a factor.

During vertical autorotation, the rotor disc is divided into three regions as illustrated in figure 3-21—the

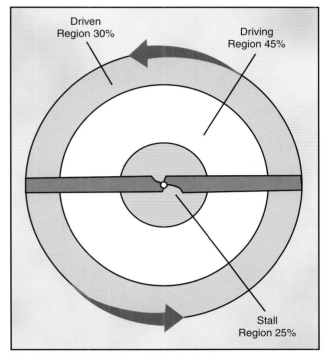

Figure 3-21. Blade regions in vertical autorotation descent.

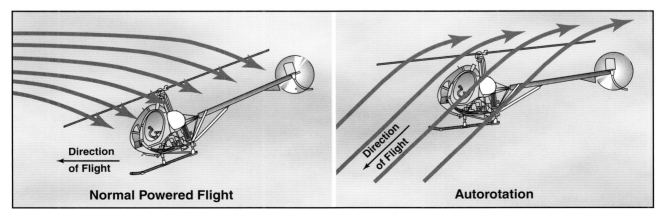

Normal Powered Flight

Autorotation

Figure 3-20. During an autorotation, the upward flow of relative wind permits the main rotor blades to rotate at their normal speed. In effect, the blades are "gliding" in their rotational plane.

driven region, the driving region, and the stall region. Figure 3-22 shows four blade sections that illustrate force vectors. Part A is the driven region, B and D are points of equilibrium, part C is the driving region, and part E is the stall region. Force vectors are different in each region because rotational relative wind is slower near the blade root and increases continually toward the blade tip. Also, blade twist gives a more positive angle of attack in the driving region than in the driven region. The combination of the inflow up through the rotor with rotational relative wind produces different combinations of aerodynamic force at every point along the blade.

The driven region, also called the propeller region, is nearest the blade tips. Normally, it consists of about 30

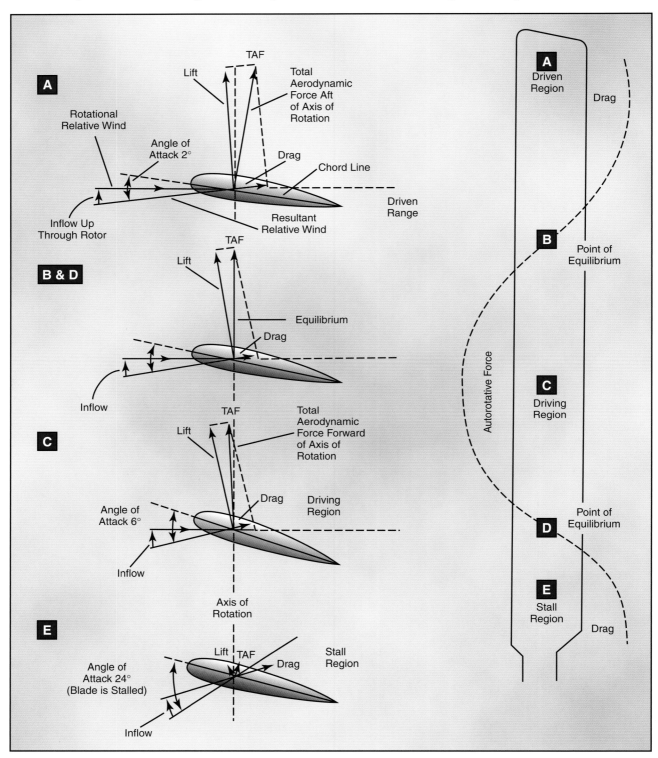

Figure 3-22. Force vectors in vertical autorotation descent.

percent of the radius. In the driven region, part A of figure 3-22, the total aerodynamic force acts behind the axis of rotation, resulting in a overall drag force. The driven region produces some lift, but that lift is offset by drag. The overall result is a deceleration in the rotation of the blade. The size of this region varies with the blade pitch, rate of descent, and rotor r.p.m. When changing autorotative r.p.m., blade pitch, or rate of descent, the size of the driven region in relation to the other regions also changes.

There are two points of equilibrium on the blade—one between the driven region and the driving region, and one between the driving region and the stall region. At points of equilibrium, total aerodynamic force is aligned with the axis of rotation. Lift and drag are produced, but the total effect produces neither acceleration nor deceleration.

The driving region, or autorotative region, normally lies between 25 to 70 percent of the blade radius. Part C of figure 3-22 shows the driving region of the blade, which produces the forces needed to turn the blades during autorotation. Total aerodynamic force in the driving region is inclined slightly forward of the axis of rotation, producing a continual acceleration force. This inclination supplies thrust, which tends to accelerate the rotation of the blade. Driving region size varies with blade pitch setting, rate of descent, and rotor r.p.m.

By controlling the size of this region you can adjust autorotative r.p.m. For example, if the collective pitch is raised, the pitch angle increases in all regions. This causes the point of equilibrium to move inboard along the blade's span, thus increasing the size of the driven region. The stall region also becomes larger while the driving region becomes smaller. Reducing the size of the driving region causes the acceleration force of the driving region and r.p.m. to decrease.

The inner 25 percent of the rotor blade is referred to as the stall region and operates above its maximum angle of attack (stall angle) causing drag which tends to slow rotation of the blade. Part E of figure 3-22 depicts the stall region.

A constant rotor r.p.m. is achieved by adjusting the collective pitch so blade acceleration forces from the driving region are balanced with the deceleration forces from the driven and stall regions.

AUTOROTATION (FORWARD FLIGHT)

Autorotative force in forward flight is produced in exactly the same manner as when the helicopter is descending vertically in still air. However, because forward speed changes the inflow of air up through the rotor disc, all three regions move outboard along the blade span on the retreating side of the disc where angle of attack is larger, as shown in figure 3-23. With lower angles of attack on the advancing side blade, more of that blade falls in the driven region. On the retreating side, more of the blade is in the stall region. A small section near the root experiences a reversed flow, therefore the size of the driven region on the retreating side is reduced.

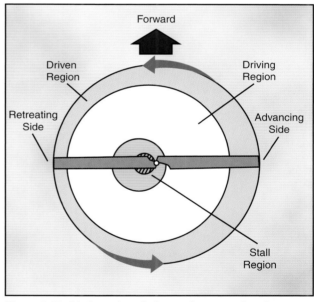

Figure 3-23. Blade regions in forward autorotation descent.

Note: In this chapter, it is assumed that the helicopter has a counterclockwise main rotor blade rotation as viewed from above. If flying a helicopter with a clockwise rotation, you will need to reverse left and right references, particularly in the areas of rotor blade pitch change, antitorque pedal movement, and tail rotor thrust.

There are four basic controls used during flight. They are the collective pitch control, the throttle, the cyclic pitch control, and the antitorque pedals.

COLLECTIVE PITCH CONTROL

The collective pitch control, located on the left side of the pilot's seat, changes the pitch angle of all main rotor blades simultaneously, or collectively, as the name implies. As the collective pitch control is raised, there is a simultaneous and equal increase in pitch angle of all main rotor blades; as it is lowered, there is a simultaneous and equal decrease in pitch angle. This is done through a series of mechanical linkages and the amount of movement in the collective lever determines the amount of blade pitch change. [Figure 4-1] An adjustable friction control helps prevent inadvertent collective pitch movement.

Changing the pitch angle on the blades changes the angle of attack on each blade. With a change in angle of attack comes a change in drag, which affects the speed or r.p.m. of the main rotor. As the pitch angle increases, angle of attack increases, drag increases, and rotor r.p.m. decreases. Decreasing pitch angle decreases both angle of attack and drag, while rotor r.p.m. increases. In order to maintain a constant rotor r.p.m., which is essential in helicopter operations, a proportionate change in power is required to compensate for the change in drag. This is accomplished with the throttle control or a correlator and/or governor, which automatically adjusts engine power.

THROTTLE CONTROL

The function of the throttle is to regulate engine r.p.m. If the correlator or governor system does not maintain the desired r.p.m. when the collective is raised or lowered, or if those systems are not installed, the throttle

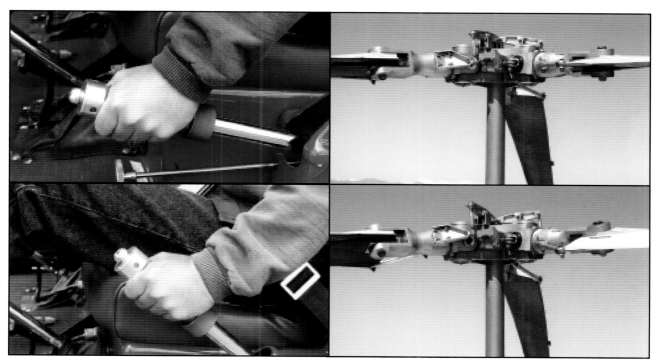

Figure 4-1. Raising the collective pitch control increases the pitch angle the same amount on all blades.

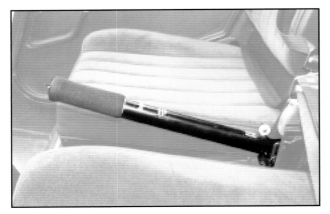

Figure 4-2. A twist grip throttle is usually mounted on the end of the collective lever. Some turbine helicopters have the throttles mounted on the overhead panel or on the floor in the cockpit.

has to be moved manually with the twist grip in order to maintain r.p.m. Twisting the throttle outboard increases r.p.m.; twisting it inboard decreases r.p.m. [Figure 4-2]

COLLECTIVE PITCH / THROTTLE COORDINATION

When the collective pitch is raised, the load on the engine is increased in order to maintain desired r.p.m. The load is measured by a manifold pressure gauge in piston helicopters or by a torque gauge in turbine helicopters.

In piston helicopters, the collective pitch is the primary control for manifold pressure, and the throttle is the primary control for r.p.m. However, the collective pitch control also influences r.p.m., and the throttle also influences manifold pressure; therefore, each is considered to be a secondary control of the other's function. Both the tachometer (r.p.m. indicator) and the manifold pressure gauge must be analyzed to determine which control to use. Figure 4-3 illustrates this relationship.

If Manifold Pressure is	and R.P.M. is	Solution
Low	Low	Increasing the throttle increases manifold pressure and r.p.m.
High	Low	Lowering the collective pitch decreases manifold pressure and increases r.p.m.
Low	High	Raising the collective pitch increases manifold pressure and decreases r.p.m.
High	High	Reducing the throttle decreases manifold pressure and r.p.m.

Figure 4-3. Relationship between manifold pressure, r.p.m., collective, and throttle.

CORRELATOR / GOVERNOR

A correlator is a mechanical connection between the collective lever and the engine throttle. When the collective lever is raised, power is automatically increased and when lowered, power is decreased. This system maintains r.p.m. close to the desired value, but still requires adjustment of the throttle for fine tuning.

A governor is a sensing device that senses rotor and engine r.p.m. and makes the necessary adjustments in order to keep rotor r.p.m. constant. In normal operations, once the rotor r.p.m. is set, the governor keeps the r.p.m. constant, and there is no need to make any throttle adjustments. Governors are common on all turbine helicopters and used on some piston powered helicopters.

Some helicopters do not have correlators or governors and require coordination of all collective and throttle movements. When the collective is raised, the throttle must be increased; when the collective is lowered, the throttle must be decreased. As with any aircraft control, large adjustments of either collective pitch or throttle should be avoided. All corrections should be made through the use of smooth pressure.

CYCLIC PITCH CONTROL

The cyclic pitch control tilts the main rotor disc by changing the pitch angle of the rotor blades in their cycle of rotation. When the main rotor disc is tilted, the horizontal component of lift moves the helicopter in the direction of tilt. [Figure 4-4]

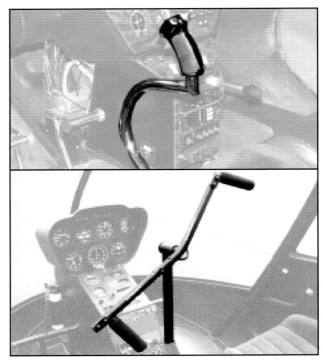

Figure 4-4. The cyclic pitch control may be mounted vertically between the pilot's knees or on a teetering bar from a single cyclic located in the center of the helicopter. The cyclic can pivot in all directions.

The rotor disc tilts in the direction that pressure is applied to the cyclic pitch control. If the cyclic is moved forward, the rotor disc tilts forward; if the cyclic is moved aft, the disc tilts aft, and so on. Because the rotor disc acts like a gyro, the mechanical linkages for the cyclic control rods are rigged in such a way that they decrease the pitch angle of the rotor blade approximately 90° before it reaches the direction of cyclic displacement, and increase the pitch angle of the rotor blade approximately 90° after it passes the direction of displacement. An increase in pitch angle increases angle of attack; a decrease in pitch angle decreases angle of attack. For example, if the cyclic is moved forward, the angle of attack decreases as the rotor blade passes the right side of the helicopter and increases on the left side. This results in maximum downward deflection of the rotor blade in front of the helicopter and maximum upward deflection behind it, causing the rotor disc to tilt forward.

ANTITORQUE PEDALS

The antitorque pedals, located on the cabin floor by the pilot's feet, control the pitch, and therefore the thrust, of the tail rotor blades. [Figure 4-5] . The main purpose of the tail rotor is to counteract the torque effect of the main rotor. Since torque varies with changes in power, the tail rotor thrust must also be varied. The pedals are connected to the pitch change mechanism on the tail rotor gearbox and allow the pitch angle on the tail rotor blades to be increased or decreased.

Figure 4-5. Antitorque pedals compensate for changes in torque and control heading in a hover.

HEADING CONTROL

Besides counteracting torque of the main rotor, the tail rotor is also used to control the heading of the helicopter while hovering or when making hovering turns. Hovering turns are commonly referred to as "pedal turns."

In forward flight, the antitorque pedals are not used to control the heading of the helicopter, except during portions of crosswind takeoffs and approaches. Instead they are used to compensate for torque to put the helicopter in longitudinal trim so that coordinated flight can be maintained. The cyclic control is used to change heading by making a turn to the desired direction.

The thrust of the tail rotor depends on the pitch angle of the tail rotor blades. This pitch angle can be positive, negative, or zero. A positive pitch angle tends to move the tail to the right. A negative pitch angle moves the tail to the left, while no thrust is produced with a zero pitch angle.

With the right pedal moved forward of the neutral position, the tail rotor either has a negative pitch angle or a small positive pitch angle. The farther it is forward, the larger the negative pitch angle. The nearer it is to neutral, the more positive the pitch angle, and somewhere in between, it has a zero pitch angle. As the left pedal is moved forward of the neutral position, the positive pitch angle of the tail rotor increases until it becomes maximum with full forward displacement of the left pedal.

If the tail rotor has a negative pitch angle, tail rotor thrust is working in the same direction as the torque of the main rotor. With a small positive pitch angle, the tail rotor does not produce sufficient thrust to overcome the torque effect of the main rotor during cruise flight. Therefore, if the right pedal is displaced forward of neutral during cruising flight, the tail rotor thrust does not overcome the torque effect, and the nose yaws to the right. [Figure 4-6]

With the antitorque pedals in the neutral position, the tail rotor has a medium positive pitch angle. In medium positive pitch, the tail rotor thrust approximately equals the torque of the main rotor during cruise flight, so the helicopter maintains a constant heading in level flight.

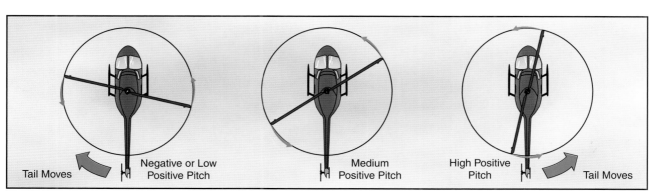

Figure 4-6. Tail rotor pitch angle and thrust in relation to pedal positions during cruising flight.

If the left pedal is in a forward position, the tail rotor has a high positive pitch position. In this position, tail rotor thrust exceeds the thrust needed to overcome torque effect during cruising flight so the helicopter yaws to the left.

The above explanation is based on cruise power and airspeed. Since the amount of torque is dependent on the amount of engine power being supplied to the main rotor, the relative positions of the pedals required to counteract torque depend upon the amount of power being used at any time. In general, the less power being used, the greater the requirement for forward displacement of the right pedal; the greater the power, the greater the forward displacement of the left pedal.

The maximum positive pitch angle of the tail rotor is generally somewhat greater than the maximum negative pitch angle available. This is because the primary purpose of the tail rotor is to counteract the torque of the main rotor. The capability for tail rotors to produce thrust to the left (negative pitch angle) is necessary, because during autorotation the drag of the transmission tends to yaw the nose to the left, or in the same direction the main rotor is turning.

By knowing the various systems on a helicopter, you will be able to more easily recognize potential problems, and if a problem arises, you will have a better understanding of what to do to correct the situation.

ENGINES

The two most common types of engines used in helicopters are the reciprocating engine and the turbine engine. Reciprocating engines, also called piston engines, are generally used in smaller helicopters. Most training helicopters use reciprocating engines because they are relatively simple and inexpensive to operate. Turbine engines are more powerful and are used in a wide variety of helicopters. They produce a tremendous amount of power for their size but are generally more expensive to operate.

RECIPROCATING ENGINE

The reciprocating engine consists of a series of pistons connected to a rotating crankshaft. As the pistons move up and down, the crankshaft rotates. The reciprocating engine gets its name from the back-and-forth movement of its internal parts. The four-stroke engine is the most common type, and refers to the four different cycles the engine undergoes to produce power. [Figure 5-1]

When the piston moves away from the cylinder head on the intake stroke, the intake valve opens and a mixture of fuel and air is drawn into the combustion chamber. As the cylinder moves back towards the cylinder head, the intake valve closes, and the fuel/air mixture is compressed. When compression is nearly complete, the spark plugs fire and the compressed mixture is ignited to begin the power stroke. The rapidly expanding gases from the controlled burning of the fuel/air mixture drive the piston away from the cylinder head, thus providing power to rotate the crankshaft. The piston then moves back toward the cylinder head on the exhaust stroke where the burned gasses are expelled through the opened exhaust valve.

Even when the engine is operated at a fairly low speed, the four-stroke cycle takes place several hundred times each minute. In a four-cylinder engine, each cylinder operates on a different stroke. Continuous rotation of a crankshaft is maintained by the precise timing of the power strokes in each cylinder.

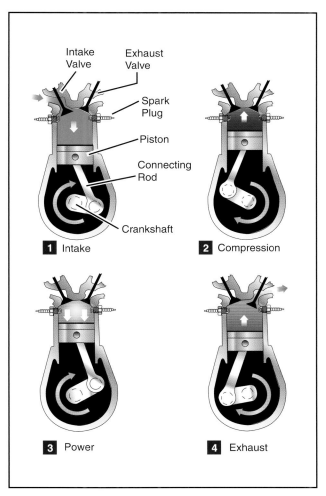

Figure 5-1. The arrows in this illustration indicate the direction of motion of the crankshaft and piston during the four-stroke cycle.

TURBINE ENGINE

The gas turbine engine mounted on most helicopters is made up of a compressor, combustion chamber, turbine, and gearbox assembly. The compressor compresses the air, which is then fed into the combustion chamber where atomized fuel is injected into it. The fuel/air mixture is ignited and allowed to expand. This combustion gas is then forced through a series of turbine wheels causing them to turn. These turbine wheels provide power to both the engine compressor and the main rotor system through an output shaft. The

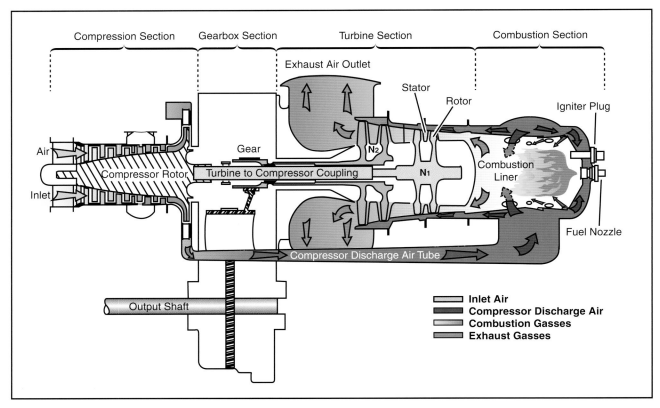

Figure 5-2. Many helicopters use a turboshaft engine to drive the main transmission and rotor systems. The main difference between a turboshaft and a turbojet engine is that most of the energy produced by the expanding gases is used to drive a turbine rather than producing thrust through the expulsion of exhaust gases.

combustion gas is finally expelled through an exhaust outlet. [Figure 5-2]

COMPRESSOR

The compressor may consist of an axial compressor, a centrifugal compressor, or both. An axial compressor consists of two main elements, the rotor and the stator. The rotor consists of a number of blades fixed on a rotating spindle and resembles a fan. As the rotor turns, air is drawn rearwards. Stator vanes are arranged in fixed rows between the rotor blades and act as a diffuser at each stage to decrease air velocity and increase air pressure. There may be a number of rows of rotor blades and stator vanes. Each row constitutes a pressure stage, and the number of stages depends on the amount of air and pressure rise required for the particular engine.

A centrifugal compressor consists of an impeller, diffuser, and a manifold. The impeller, which is a forged disc with integral blades, rotates at a high speed to draw air in and expel it at an accelerated rate. The air then passes through the diffuser which slows the air down. When the velocity of the air is slowed, static pressure increases, resulting in compressed, high-pressure air. The high pressure air then passes through the compressor manifold where it is distributed to the combustion chamber.

COMBUSTION CHAMBER

Unlike a piston engine, the combustion in a turbine engine is continuous. An igniter plug serves only to ignite the fuel/air mixture when starting the engine. Once the fuel/air mixture is ignited, it will continue to burn as long as the fuel/air mixture continues to be present. If there is an interruption of fuel, air, or both, combustion ceases. This is known as a "flame-out," and the engine has to be restarted or re-lit. Some helicopters are equipped with auto-relight, which automatically activates the igniters to start combustion if the engine flames out.

TURBINE

The turbine section consists of a series of turbine wheels that are used to drive the compressor section and the rotor system. The first stage, which is usually referred to as the gas producer or N_1 may consist of one or more turbine wheels. This stage drives the components necessary to complete the turbine cycle making the engine self-sustaining. Common components driven by the N_1 stage are the compressor, oil pump, and fuel pump. The second stage, which may also consist of one or more wheels, is dedicated to driving the main rotor system and accessories from the engine gearbox. This is referred to as the power turbine (N_2 or N_r).

If the first and second stage turbines are mechanically coupled to each other, the system is said to be a direct-drive engine or fixed turbine. These engines share a common shaft, which means the first and second stage turbines, and thus the compressor and output shaft, are connected.

On most turbine assemblies used in helicopters, the first stage and second stage turbines are not mechanically connected to each other. Rather, they are mounted on independent shafts and can turn freely with respect to each other. This is referred to as a "free turbine." When the engine is running, the combustion gases pass through the first stage turbine to drive the compressor rotor, and then past the independent second stage turbine, which turns the gearbox to drive the output shaft.

TRANSMISSION SYSTEM

The transmission system transfers power from the engine to the main rotor, tail rotor, and other accessories. The main components of the transmission system are the main rotor transmission, tail rotor drive system, clutch, and freewheeling unit. Helicopter transmissions are normally lubricated and cooled with their own oil supply. A sight gauge is provided to check the oil level. Some transmissions have **chip detectors** located in the sump. These detectors are wired to warning lights located on the pilot's instrument panel that illuminate in the event of an internal problem.

MAIN ROTOR TRANSMISSION

The primary purpose of the main rotor transmission is to reduce engine output r.p.m. to optimum rotor r.p.m. This reduction is different for the various helicopters, but as an example, suppose the engine r.p.m. of a specific helicopter is 2,700. To achieve a rotor speed of 450 r.p.m. would require a 6 to 1 reduction. A 9 to 1 reduction would mean the rotor would turn at 300 r.p.m.

Most helicopters use a dual-needle tachometer to show both engine and rotor r.p.m. or a percentage of engine and rotor r.p.m. The rotor r.p.m. needle normally is used only during clutch engagement to monitor rotor acceleration, and in autorotation to maintain r.p.m. within prescribed limits. [Figure 5-3]

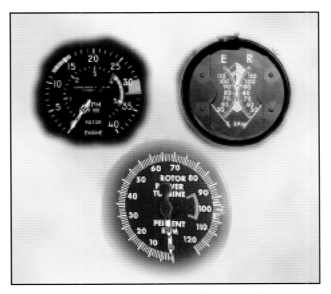

Figure 5-3. There are various types of dual-needle tachometers, however, when the needles are superimposed or married, the ratio of the engine r.p.m. is the same as the gear reduction ratio.

In helicopters with horizontally mounted engines, another purpose of the main rotor transmission is to change the axis of rotation from the horizontal axis of the engine to the vertical axis of the rotor shaft.

TAIL ROTOR DRIVE SYSTEM

The tail rotor drive system consists of a tail rotor drive shaft powered from the main transmission and a tail rotor transmission mounted at the end of the tail boom. The drive shaft may consist of one long shaft or a series of shorter shafts connected at both ends with flexible couplings. This allows the drive shaft to flex with the tail boom. The tail rotor transmission provides a right angle drive for the tail rotor and may also include gearing to adjust the output to optimum tail rotor r.p.m. [Figure 5-4]

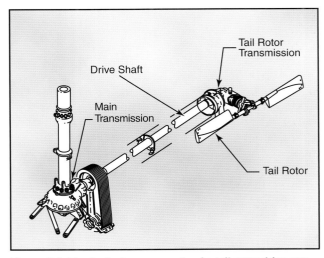

Figure 5-4. The typical components of a tail rotor drive system are shown here.

Chip Detector—A chip detector is a warning device that alerts you to any abnormal wear in a transmission or engine. It consists of a magnetic plug located within the transmission. The magnet attracts any ferrous metal particles that have come loose from the bearings or other transmission parts. Most chip detectors send a signal to lights located on the instrument panel that illuminate when ferrous metal particles are picked up.

CLUTCH

In a conventional airplane, the engine and propeller are permanently connected. However, in a helicopter there is a different relationship between the engine and the rotor. Because of the greater weight of a rotor in relation to the power of the engine, as compared to the weight of a propeller and the power in an airplane, the rotor must be disconnected from the engine when you engage the starter. A clutch allows the engine to be started and then gradually pick up the load of the rotor.

On free turbine engines, no clutch is required, as the gas producer turbine is essentially disconnected from the power turbine. When the engine is started, there is little resistance from the power turbine. This enables the gas producer turbine to accelerate to normal idle speed without the load of the transmission and rotor system dragging it down. As the gas pressure increases through the power turbine, the rotor blades begin to turn, slowly at first and then gradually accelerate to normal operating r.p.m.

On reciprocating helicopters, the two main types of clutches are the centrifugal clutch and the belt drive clutch.

CENTRIFUGAL CLUTCH

The centrifugal clutch is made up of an inner assembly and a outer drum. The inner assembly, which is connected to the engine driveshaft, consists of shoes lined with material similar to automotive brake linings. At low engine speeds, springs hold the shoes in, so there is no contact with the outer drum, which is attached to the transmission input shaft. As engine speed increases, centrifugal force causes the clutch shoes to move outward and begin sliding against the outer drum. The transmission input shaft begins to rotate, causing the rotor to turn, slowly at first, but increasing as the friction increases between the clutch shoes and transmission drum. As rotor speed increases, the rotor tachometer needle shows an increase by moving toward the engine tachometer needle. When the two needles are superimposed, the engine and the rotor are synchronized, indicating the clutch is fully engaged and there is no further slippage of the clutch shoes.

BELT DRIVE CLUTCH

Some helicopters utilize a belt drive to transmit power from the engine to the transmission. A belt drive consists of a lower pulley attached to the engine, an upper pulley attached to the transmission input shaft, a belt or a series of V-belts, and some means of applying tension to the belts. The belts fit loosely over the upper and lower pulley when there is no tension on the belts. This allows the engine to be started without any load from the transmission. Once the engine is running, tension on the belts is gradually increased. When the rotor and engine tachometer needles are superimposed, the rotor and the engine are synchronized, and the clutch is then fully engaged.

Advantages of this system include vibration isolation, simple maintenance, and the ability to start and warm up the engine without engaging the rotor.

FREEWHEELING UNIT

Since lift in a helicopter is provided by rotating airfoils, these airfoils must be free to rotate if the engine fails. The freewheeling unit automatically disengages the engine from the main rotor when engine r.p.m. is less than main rotor r.p.m. This allows the main rotor to continue turning at normal in-flight speeds. The most common freewheeling unit assembly consists of a one-way sprag clutch located between the engine and main rotor transmission. This is usually in the upper pulley in a piston helicopter or mounted on the engine gearbox in a turbine helicopter. When the engine is driving the rotor, inclined surfaces in the spray clutch force rollers against an outer drum. This prevents the engine from exceeding transmission r.p.m. If the engine fails, the rollers move inward, allowing the outer drum to exceed the speed of the inner portion. The transmission can then exceed the speed of the engine. In this condition, engine speed is less than that of the drive system, and the helicopter is in an autorotative state.

MAIN ROTOR SYSTEM

Main rotor systems are classified according to how the main rotor blades move relative to the main rotor hub. As was described in Chapter 1—Introduction to the Helicopter, there are three basic classifications: fully articulated, semirigid, or rigid. Some modern rotor systems use a combination of these types.

FULLY ARTICULATED ROTOR SYSTEM

In a fully articulated rotor system, each rotor blade is attached to the rotor hub through a series of hinges, which allow the blade to move independently of the others. These rotor systems usually have three or more blades. [Figure 5-5]

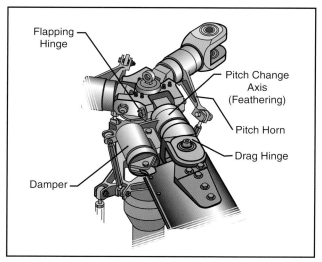

Figure 5-5. Each blade of a fully articulated rotor system can flap, drag, and feather independently of the other blades.

The horizontal hinge, called the flapping hinge, allows the blade to move up and down. This movement is called flapping and is designed to compensate for dissymetry of lift. The flapping hinge may be located at varying distances from the rotor hub, and there may be more than one hinge.

The vertical hinge, called the lead-lag or drag hinge, allows the blade to move back and forth. This movement is called lead-lag, dragging, or hunting. Dampers are usually used to prevent excess back and forth movement around the drag hinge. The purpose of the drag hinge and dampers is to compensate for the acceleration and deceleration caused by Coriolis Effect.

Each blade can also be feathered, that is, rotated around its spanwise axis. Feathering the blade means changing the pitch angle of the blade. By changing the pitch angle of the blades you can control the thrust and direction of the main rotor disc.

SEMIRIGID ROTOR SYSTEM

A semirigid rotor system is usually composed of two blades which are rigidly mounted to the main rotor hub. The main rotor hub is free to tilt with respect to the main rotor shaft on what is known as a teetering hinge. This allows the blades to flap together as a unit. As one blade flaps up, the other flaps down. Since there is no vertical drag hinge, lead-lag forces are absorbed through blade bending. [Figure 5-6]

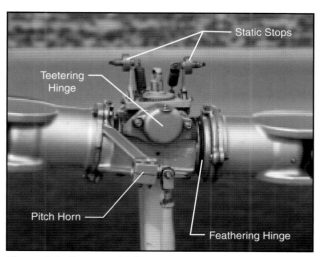

Figure 5-6. On a semirigid rotor system, a teetering hinge allows the rotor hub and blades to flap as a unit. A static flapping stop located above the hub prevents excess rocking when the blades are stopped. As the blades begin to turn, centrifugal force pulls the static stops out of the way.

RIGID ROTOR SYSTEM

In a rigid rotor system, the blades, hub, and mast are rigid with respect to each other. There are no vertical or horizontal hinges so the blades cannot flap or drag, but

they can be feathered. Flapping and lead/lag forces are absorbed by blade bending.

COMBINATION ROTOR SYSTEMS

Modern rotor systems may use the combined principles of the rotor systems mentioned above. Some rotor hubs incorporate a flexible hub, which allows for blade bending (flexing) without the need for bearings or hinges. These systems, called flextures, are usually constructed from composite material. Elastomeric bearings may also be used in place of conventional roller bearings. Elastomeric bearings are bearings constructed from a rubber type material and have limited movement that is perfectly suited for helicopter applications. Flextures and elastomeric bearings require no lubrication and, therefore, require less maintenance. They also absorb vibration, which means less fatigue and longer service life for the helicopter components. [Figure 5-7]

Figure 5-7. Rotor systems, such as Eurocopter's Starflex or Bell's soft-in-plane, use composite material and elastomeric bearings to reduce complexity and maintenance and, thereby, increase reliability.

SWASH PLATE ASSEMBLY

The purpose of the swash plate is to transmit control inputs from the collective and cyclic controls to the main rotor blades. It consists of two main parts: the stationary swash plate and the rotating swash plate. [Figure 5-8] The stationary swash plate is mounted around the main rotor mast and connected to the cyclic and collective controls by a series of pushrods. It is restrained from rotating but is able to tilt in all directions and move vertically. The rotating swash plate is mounted to the stationary swash plate by means of a bearing and is allowed to rotate with the main rotor mast. Both swash plates tilt and slide up and down as one unit. The rotating swash plate is connected to the pitch horns by the pitch links.

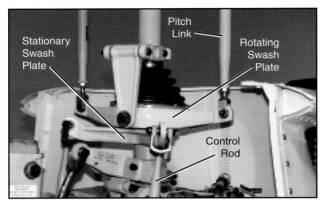

Figure 5-8. Collective and cyclic control inputs are transmitted to the stationary swash plate by control rods causing it to tilt or to slide vertically. The pitch links attached from the rotating swash plate to the pitch horns on the rotor hub transmit these movements to the blades.

FUEL SYSTEMS

The fuel system in a helicopter is made up of two groups of components: the fuel supply system and the engine fuel control system.

FUEL SUPPLY SYSTEM

The supply system consists of a fuel tank or tanks, fuel quantity gauges, a shut-off valve, fuel filter, a fuel line to the engine, and possibly a primer and fuel pumps. [Figure 5-9]

The fuel tanks are usually mounted to the airframe as close as possible to the center of gravity. This way, as fuel is burned off, there is a negligible effect on the center of gravity. A drain valve located on the bottom of the fuel tank allows the pilot to drain water and sediment that may have collected in the tank. A fuel vent prevents the formation of a vacuum in the tank, and an overflow drain allows for fuel to expand without rupturing the tank. A fuel quantity gauge located on the pilot's instrument panel shows the amount of fuel measured by a sensing unit inside the tank. Some gauges show tank capacity in both gallons and pounds.

The fuel travels from the fuel tank through a shut-off valve, which provides a means to completely stop fuel

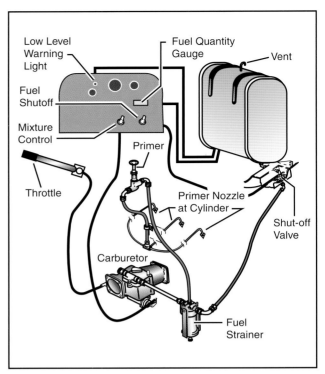

Figure 5-9. A typical gravity feed fuel system, in a helicopter with a reciprocating engine, contains the components shown here.

flow to the engine in the event of an emergency or fire. The shut-off valve remains in the open position for all normal operations.

Most non-gravity feed fuel systems contain both an electric pump and a mechanical engine driven pump. The electrical pump is used to maintain positive fuel pressure to the engine pump and also serves as a backup in the event of mechanical pump failure. The electrical pump is controlled by a switch in the cockpit. The engine driven pump is the primary pump that supplies fuel to the engine and operates any time the engine is running.

A fuel filter removes moisture and other sediment from the fuel before it reaches the engine. These contaminants are usually heavier than fuel and settle to the bottom of the fuel filter sump where they can be drained out by the pilot.

Some fuel systems contain a small hand-operated pump called a primer. A primer allows fuel to be pumped directly into the intake port of the cylinders prior to engine start. The primer is useful in cold weather when fuel in the carburetor is difficult to vaporize.

ENGINE FUEL CONTROL SYSTEM

The purpose of the fuel control system is to bring outside air into the engine, mix it with fuel in the proper proportion, and deliver it to the combustion chamber.

RECIPROCATING ENGINES
Fuel is delivered to the cylinders by either a carburetor or fuel injection system.

CARBURETOR
In a carburetor system, air is mixed with vaporized fuel as it passes through a venturi in the carburetor. The metered fuel/air mixture is then delivered to the cylinder intake.

Carburetors are calibrated at sea level, and the correct fuel-to-air mixture ratio is established at that altitude with the mixture control set in the FULL RICH position. However, as altitude increases, the density of air entering the carburetor decreases while the density of the fuel remains the same. This means that at higher altitudes, the mixture becomes progressively richer. To maintain the correct fuel/air mixture, you must be able to adjust the amount of fuel that is mixed with the incoming air. This is the function of the mixture control. This adjustment, often referred to as "leaning the mixture," varies from one aircraft to another. Refer to the FAA-Approved Rotorcraft Flight Manual (RFM) to determine specific procedures for your helicopter. Note that most manufacturers do not recommend leaning helicopters in-flight.

Most mixture adjustments are required during changes of altitude or during operations at airports with field elevations well above sea level. A mixture that is too rich can result in engine roughness and reduced power. The roughness normally is due to spark plug fouling from excessive carbon buildup on the plugs. This occurs because the excessively rich mixture lowers the temperature inside the cylinder, inhibiting complete combustion of the fuel. This condition may occur during the pretakeoff runup at high elevation airports and during climbs or cruise flight at high altitudes. Usually, you can correct the problem by leaning the mixture according to RFM instructions.

If you fail to enrich the mixture during a descent from high altitude, it normally becomes too lean. High engine temperatures can cause excessive engine wear or even failure. The best way to avoid this type of situation is to monitor the engine temperature gauges regularly and follow the manufacturer's guidelines for maintaining the proper mixture.

CARBURETOR ICE
The effect of fuel vaporization and decreasing air pressure in the venturi causes a sharp drop in temperature in the carburetor. If the air is moist, the water vapor in the air may condense. When the temperature in the carburetor is at or below freezing, carburetor ice may form on internal surfaces, including the throttle valve. [Figure 5-10] Because of the sudden cooling that takes place in the carburetor, icing can occur even on warm days with temperatures as high as 38°C (100°F) and the humidity as low as 50 percent. However, it is more likely to occur when temperatures are below 21°C (70°F) and the relative humidity is above 80 percent. The likelihood of icing increases as temperature decreases down to 0°C (32°F), and as relative humidity increases. Below freezing, the possibility of carburetor icing decreases with decreasing temperatures.

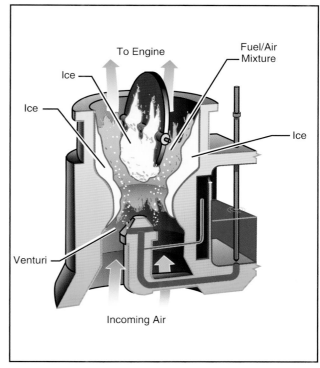

Figure 5-10. Carburetor ice reduces the size of the air passage to the engine. This restricts the flow of the fuel/air mixture, and reduces power.

Although carburetor ice can occur during any phase of flight, it is particularly dangerous when you are using reduced power, such as during a descent. You may not notice it during the descent until you try to add power.

Indications of carburetor icing are a decrease in engine r.p.m. or manifold pressure, the carburetor air temperature gauge indicating a temperature outside the safe operating range, and engine roughness. Since changes in r.p.m. or manifold pressure can occur for a number of reasons, it is best to closely check the carburetor air temperature gauge when in possible carburetor icing conditions. Carburetor air temperature gauges are marked with a yellow caution arc or green operating arcs. You should refer to the FAA-Approved Rotorcraft Flight Manual for the specific procedure as to when and how to apply carburetor heat. However, in most cases, you should keep the needle out of the yellow arc or in the green arc. This is accomplished by using a carburetor heat system, which eliminates the ice by routing air across a heat source, such as an exhaust manifold, before it enters the carburetor. [Figure 5-11].

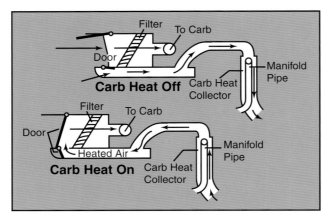

Figure 5-11. When you turn the carburetor heat ON, normal air flow is blocked, and heated air from an alternate source flows through the filter to the carburetor.

FUEL INJECTION

In a fuel injection system, fuel and air are metered at the fuel control unit but are not mixed. The fuel is injected directly into the intake port of the cylinder where it is mixed with the air just before entering the cylinder. This system ensures a more even fuel distribution in the cylinders and better vaporization, which in turn, promotes more efficient use of fuel. Also, the fuel injection system eliminates the problem of carburetor icing and the need for a carburetor heat system.

TURBINE ENGINES

The fuel control system on the turbine engine is fairly complex, as it monitors and adjusts many different parameters on the engine. These adjustments are done automatically and no action is required of the pilot other than starting and shutting down. No mixture adjustment is necessary, and operation is fairly simple as far as the pilot is concerned. New generation fuel controls incorporate the use of a full authority digital engine control (FADEC) computer to control the engine's fuel requirements. The FADEC systems increase efficiency, reduce engine wear, and also reduce pilot workload. The FADEC usually incorporates back-up systems in the event of computer failure.

ELECTRICAL SYSTEMS

The electrical systems, in most helicopters, reflect the increased use of sophisticated avionics and other electrical accessories. More and more operations in today's flight environment are dependent on the aircraft's electrical system; however, all helicopters can be safely flown without any electrical power in the event of an electrical malfunction or emergency.

Helicopters have either a 14- or 28-volt, direct-current electrical system. On small, piston powered helicopters, electrical energy is supplied by an engine-driven alternator. These alternators have advantages over older style generators as they are lighter in weight, require lower maintenance, and maintain a uniform electrical output even at low engine r.p.m. [Figure 5-12]

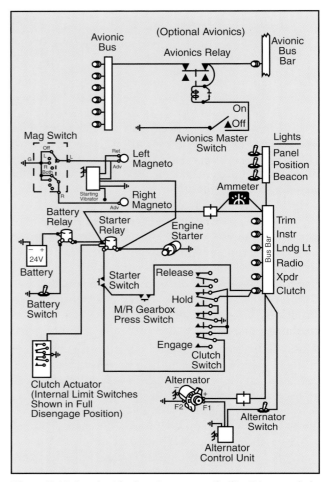

Figure 5-12. An electrical system scematic like this sample is included in most POHs. Notice that the various bus bar accessories are protected by circuit breakers. However, you should still make sure all electrical equipment is turned off before you start the engine. This protects sensitive components, particularly the radios, from damage which may be caused by random voltages generated during the starting process.

Turbine powered helicopters use a starter/generator system. The starter/generator is permanently coupled to the engine gearbox. When starting the engine, electrical power from the battery is supplied to the starter/generator, which turns the engine over. Once the engine is running, the starter/generator is driven by the engine and is then used as a generator.

Current from the alternator or generator is delivered through a voltage regulator to a bus bar. The voltage regulator maintains the constant voltage required by the electrical system by regulating the output of the alternator or generator. An over-voltage control may be incorporated to prevent excessive voltage, which may damage the electrical components. The bus bar serves to distribute the current to the various electrical components of the helicopter.

A battery is mainly used for starting the engine. In addition, it permits limited operation of electrical components, such as radios and lights, without the engine running. The battery is also a valuable source of standby or emergency electrical power in the event of alternator or generator failure.

An ammeter or loadmeter is used to monitor the electrical current within the system. The ammeter reflects current flowing to and from the battery. A charging ammeter indicates that the battery is being charged. This is normal after an engine start since the battery power used in starting is being replaced. After the battery is charged, the ammeter should stabilize near zero since the alternator or generator is supplying the electrical needs of the system. A discharging ammeter means the electrical load is exceeding the output of the alternator or generator, and the battery is helping to supply electrical power. This may mean the alternator or generator is malfunctioning, or the electrical load is excessive. A loadmeter displays the load placed on the alternator or generator by the electrical equipment. The RFM for a particular helicopter shows the normal load to expect. Loss of the alternator or generator causes the loadmeter to indicate zero.

Electrical switches are used to select electrical components. Power may be supplied directly to the component or to a relay, which in turn provides power to the component. Relays are used when high current and/or heavy electrical cables are required for a particular component, which may exceed the capacity of the switch.

Circuit breakers or fuses are used to protect various electrical components from overload. A circuit breaker pops out when its respective component is overloaded. The circuit breaker may be reset by pushing it back in, unless a short or the overload still exists. In this case, the circuit breaker continues to pop, indicating an electrical malfunction. A fuse simply burns out when it is overloaded and needs to be replaced. Manufacturers usually provide a holder for spare fuses in the event one has to be replaced in flight. Caution lights on the instrument panel may be installed to show the malfunction of an electrical component.

HYDRAULICS

Most helicopters, other than smaller piston powered helicopters, incorporate the use of hydraulic actuators to overcome high control forces. [Figure 5-13] A typical hydraulic system consists of actuators, also called

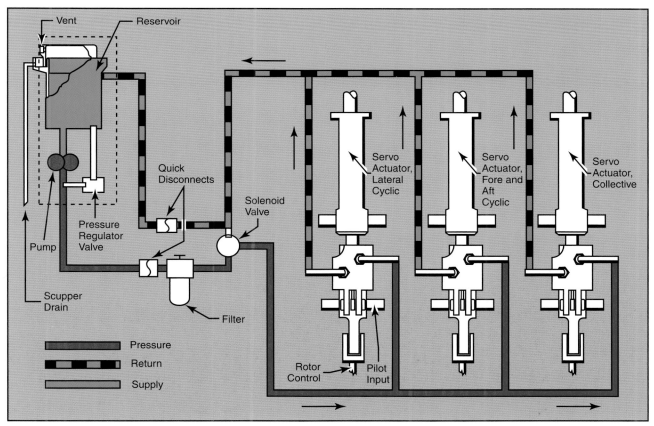

Figure 5-13. A typical hydraulic system for helicopters in the light to medium range is shown here.

servos, on each flight control, a pump which is usually driven by the main rotor gearbox, and a reservoir to store the hydraulic fluid. A switch in the cockpit can turn the system off, although it is left on under normal conditions. A pressure indicator in the cockpit may also be installed to monitor the system.

When you make a control input, the servo is activated and provides an assisting force to move the respective flight control, thus lightening the force required by the pilot. These boosted flight controls ease pilot workload and fatigue. In the event of hydraulic system failure, you are still able to control the helicopter, but the control forces will be very heavy.

In those helicopters where the control forces are so high that they cannot be moved without hydraulic assistance, two or more independent hydraulic systems may be installed. Some helicopters use hydraulic accumulators to store pressure, which can be used for a short period of time in an emergency if the hydraulic pump fails. This gives you enough time to land the helicopter with normal control

STABILITY AUGMENTATIONS SYSTEMS

Some helicopters incorporate stability augmentations systems (SAS) to aid in stabilizing the helicopter in flight and in a hover. The simplest of these systems is a force trim system, which uses a magnetic clutch and springs to hold the cyclic control in the position where it was released. More advanced systems use electric servos that actually move the flight controls. These servos receive control commands from a computer that senses helicopter attitude. Other inputs, such as heading, speed, altitude, and navigation information may be supplied to the computer to form a complete autopilot system. The SAS may be overridden or disconnected by the pilot at any time.

Stability augmentation systems reduce pilot workload by improving basic aircraft control harmony and decreasing disturbances. These systems are very useful when you are required to perform other duties, such as sling loading and search and rescue operations.

AUTOPILOT

Helicopter autopilot systems are similar to stability augmentations systems except they have additional features. An autopilot can actually fly the helicopter and perform certain functions selected by the pilot. These functions depend on the type of autopilot and systems installed in the helicopter.

The most common functions are altitude and heading hold. Some more advanced systems include a vertical speed or indicated airspeed (IAS) hold mode, where a constant rate of climb/descent or indicated airspeed is maintained by the autopilot. Some autopilots have nav-

igation capabilities, such as VOR, ILS, and GPS intercept and tracking, which is especially useful in IFR conditions. The most advanced autopilots can fly an instrument approach to a hover without any additional pilot input once the initial functions have been selected.

The autopilot system consists of electric actuators or servos connected to the flight controls. The number and location of these servos depends on the type of system installed. A two-axis autopilot controls the helicopter in pitch and roll; one servo controls fore and aft cyclic, and another controls left and right cyclic. A three-axis autopilot has an additional servo connected to the anti-torque pedals and controls the helicopter in yaw. A four-axis system uses a fourth servo which controls the collective. These servos move the respective flight controls when they receive control commands from a central computer. This computer receives data input from the flight instruments for attitude reference and from the navigation equipment for navigation and tracking reference. An autopilot has a control panel in the cockpit that allows you to select the desired functions, as well as engage the autopilot.

For safety purposes, an automatic disengage feature is usually included which automatically disconnects the autopilot in heavy turbulence or when extreme flight attitudes are reached. Even though all autopilots can be overridden by the pilot, there is also an autopilot disengage button located on the cyclic or collective which allows you to completely disengage the autopilot without removing your hands from the controls. Because autopilot systems and installations differ from one helicopter to another, it is very important that you refer to the autopilot operating procedures located in the Rotorcraft Flight Manual.

ENVIRONMENTAL SYSTEMS

Heating and cooling for the helicopter cabin can be provided in different ways. The simplest form of cooling is ram air cooling. Air ducts in the front or sides of the helicopter are opened or closed by the pilot to let ram air into the cabin. This system is limited as it requires forward airspeed to provide airflow and also

---

VOR—Ground-based navigation system consisting of very high frequency omnidirectional range (VOR) stations which provide course guidance.

ILS (Instrument Landing System)—A precision instrument approach system, which normally consists of the following electronic components and visual aids: localizer, glide slope, outer marker, and approach lights.

GPS (Global Positioning System)—A satellite-based radio positioning, navigation, and time-transfer system.

IFR (Instrument Flight Rules)—Rules that govern the procedure for conducting flight in weather conditions below VFR weather minimums. The term IFR also is used to define weather conditions and the type of flight plan under which an aircraft is operating.

depends on the temperature of the outside air. Air conditioning provides better cooling but it is more complex and weighs more than a ram air system.

Piston powered helicopters use a heat exchanger shroud around the exhaust manifold to provide cabin heat. Outside air is piped to the shroud and the hot exhaust manifold heats the air, which is then blown into the cockpit. This warm air is heated by the exhaust manifold but is not exhaust gas. Turbine helicopters use a bleed air system for heat. Bleed air is hot, compressed, discharge air from the engine compressor. Hot air is ducted from the compressor to the helicopter cabin through a pilot-controlled, bleed air valve.

ANTI-ICING SYSTEMS

Most anti-icing equipment installed on small helicopters is limited to engine intake anti-ice and pitot heat systems.

The anti-icing system found on most turbine-powered helicopters uses engine bleed air. The bleed air flows through the inlet guide vanes to prevent ice formation on the hollow vanes. A pilot-controlled, electrically operated valve on the compressor controls the air flow. The pitot heat system uses an electrical element to heat the pitot tube, thus melting or preventing ice formation.

Airframe and rotor anti-icing may be found on some larger helicopters, but it is not common due to the complexity, expense, and weight of such systems. The leading edges of rotors may be heated with bleed air or electrical elements to prevent ice formation. Balance and control problems might arise if ice is allowed to form unevenly on the blades. Research is being done on lightweight ice-phobic (anti-icing) materials or coatings. These materials placed in strategic areas could significantly reduce ice formation and improve performance.

CHAPTER 6

Helicopter

Rotorcraft Flight Manual

Title 14 of the Code of Federal Regulations (14 CFR) part 91 requires that pilots comply with the operating limitations specified in approved rotorcraft flight manuals, markings, and placards. Originally, flight manuals were often characterized by a lack of essential information and followed whatever format and content the manufacturer felt was appropriate. This changed with the acceptance of the General Aviation Manufacturers Association's (GAMA) *Specification for Pilot's Operating Handbook*, which established a standardized format for all general aviation airplane and rotorcraft flight manuals. The term "Pilot's Operating Handbook (POH)" is often used in place of "Rotorcraft Flight Manual (RFM)." However, if "Pilot's Operating Handbook" is used as the main title instead of "Rotorcraft Flight Manual," a statement must be included on the title page indicating that the document is the FAA-Approved Rotorcraft Flight Manual. [Figure 6-1]

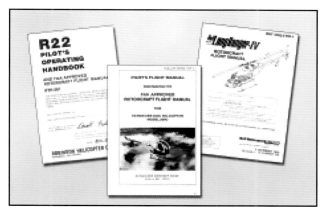

Figure 6-1. The Rotorcraft Flight Manual is a regulatory document in terms of the maneuvers, procedures, and operating limitations described therein.

Besides the preliminary pages, an FAA-Approved Rotorcraft Flight Manual may contain as many as ten sections. These sections are: General Information; Operating Limitations; Emergency Procedures; Normal Procedures; Performance; Weight and Balance; Aircraft and Systems Description; Handling, Servicing, and Maintenance; and Supplements. Manufacturers have the option of including a tenth section on Safety and Operational Tips and an alphabetical index at the end of the handbook.

PRELIMINARY PAGES

While rotorcraft flight manuals may appear similar for the same make and model of aircraft, each flight man-

ual is unique since it contains specific information about a particular aircraft, such as the equipment installed, and weight and balance information. Therefore, manufacturers are required to include the serial number and registration on the title page to identify the aircraft to which the flight manual belongs. If a flight manual does not indicate a specific aircraft registration and serial number, it is limited to general study purposes only.

Most manufacturers include a table of contents, which identifies the order of the entire manual by section number and title. Usually, each section also contains its own table of contents. Page numbers reflect the section you are reading, 1-1, 2-1, 3-1, and so on. If the flight manual is published in looseleaf form, each section is usually marked with a divider tab indicating the section number or title, or both. The Emergency Procedures section may have a red tab for quick identification and reference.

GENERAL INFORMATION

The General Information section provides the basic descriptive information on the rotorcraft and the powerplant. In some manuals there is a three-view drawing of the rotorcraft that provides the dimensions of various components, including the overall length and width, and the diameter of the rotor systems. This is a good place to quickly familiarize yourself with the aircraft.

You can find definitions, abbreviations, explanations of symbology, and some of the terminology used in the manual at the end of this section. At the option of the manufacturer, metric and other conversion tables may also be included.

OPERATING LIMITATIONS

The Operating Limitations section contains only those limitations required by regulation or that are necessary for the safe operation of the rotorcraft, powerplant, systems, and equipment. It includes operating limitations, instrument markings, color coding, and basic placards. Some of the areas included are: airspeed, altitude, rotor, and powerplant limitations, including fuel and oil requirements; weight and loading distribution; and flight limitations.

AIRSPEED LIMITATIONS

Airspeed limitations are shown on the airspeed indicator by color coding and on placards or graphs in the aircraft. A red line on the airspeed indicator shows the airspeed limit beyond which structural damage could occur. This is called the never exceed speed, or V$_{NE}$. The normal operating speed range is depicted by a green arc. A blue line is sometimes added to show the maximum safe autorotation speed. [Figure 6-2]

Figure 6-2. Typical airspeed indicator limitations and markings.

ALTITUDE LIMITATIONS

If the rotorcraft has a maximum operating density altitude, it is indicated in this section of the flight manual. Sometimes the maximum altitude varies based on different gross weights.

ROTOR LIMITATIONS

Low rotor r.p.m. does not produce sufficient lift, and high r.p.m. may cause structural damage, therefore rotor r.p.m. limitations have minimum and maximum values. A green arc depicts the normal operating range with red lines showing the minimum and maximum limits. [Figure 6-3]

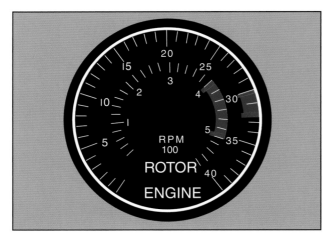

Figure 6-3. Markings on a typical dual-needle tachometer in a reciprocating-engine helicopter. The outer band shows the limits of the superimposed needles when the engine is turning the rotor. The inner band indicates the power-off limits.

There are two different rotor r.p.m. limitations: power-on and power-off. Power-on limitations apply anytime the engine is turning the rotor and is depicted by a fairly narrow green band. A yellow arc may be included to show a transition range, which means that operation within this range is limited. Power-off limitations apply anytime the engine is not turning the rotor, such as when in an autorotation. In this case, the green arc is wider than the power-on arc, indicating a larger operating range.

POWERPLANT LIMITATIONS

The Powerplant Limitations area describes operating limitations on the rotorcraft's engine including such items as r.p.m. range, power limitations, operating temperatures, and fuel and oil requirements. Most turbine engines and some reciprocating engines have a maximum power and a maximum continuous power rating. The "maximum power" rating is the maximum power the engine can generate and is usually limited by time. The maximum power range is depicted by a yellow arc on the engine power instruments, with a red line indicating the maximum power that must not be exceeded. "Maximum continuous power" is the maximum power the engine can generate continually, and is depicted by a green arc. [Figure 6-4]

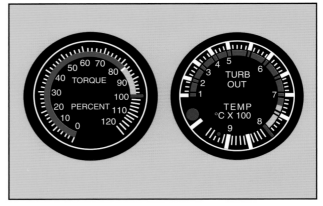

Figure 6-4. Torque and turbine outlet temperature (TOT) gauges are commonly used with turbine-powered aircraft.

Like on a torque and turbine outlet temperature gauge, the red line on a manifold pressure gauge indicates the maximum amount of power. A yellow arc on the gauge warns of pressures approaching the limit of rated power. A placard near the gauge lists the maximum readings for specific conditions. [Figure 6-5]

WEIGHT AND LOADING DISTRIBUTION

The Weight and Loading Distribution area contains the maximum certificated weights, as well as the center of gravity (CG) range. The location of the reference datum used in balance computations should also be included in this section. Weight and balance computations are not provided here, but rather in the Weight and Balance Section of the FAA-Approved Rotorcraft Flight Manual.

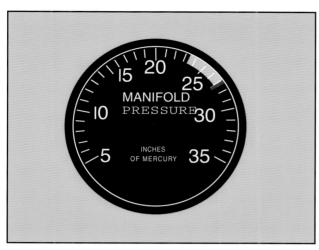

Figure 6-5. A manifold pressure gauge is commonly used with piston-powered aircraft.

FLIGHT LIMITATIONS

This area lists any maneuvers which are prohibited, such as acrobatic flight or flight into known icing conditions. If the rotorcraft can only be flown in VFR conditions, it will be noted in this area. Also included are the minimum crew requirements, and the pilot seat location, if applicable, where solo flights must be conducted.

PLACARDS

All rotorcraft generally have one or more placards displayed that have a direct and important bearing on the safe operation of the rotorcraft. These placards are located in a conspicuous place within the cabin and normally appear in the Limitations Section. Since V_{NE} changes with altitude, this placard can be found in all helicopters. [Figure 6-6]

EMERGENCY PROCEDURES

Concise checklists describing the recommended procedures and airspeeds for coping with various types of emergencies or critical situations can be found in this section. Some of the emergencies covered include: engine failure in a hover and at altitude, tail rotor failures, fires, and systems failures. The procedures for restarting an engine and for ditching in the water might also be included.

Manufacturers may first show the emergencies checklists in an abbreviated form with the order of items reflecting the sequence of action. This is followed by amplified checklists providing additional information to help you understand the procedure. To be prepared for an abnormal or emergency situation, memorize the first steps of each checklist, if not all the steps. If time permits, you can then refer to the checklist to make sure all items have been covered. (For more information on emergencies, refer to Chapter 11—Helicopter Emergencies and Chapter 21—Gyroplane Emergencies.)

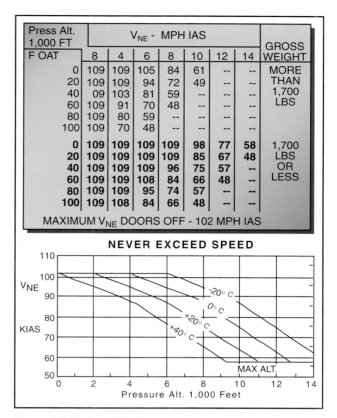

Press Alt. 1,000 FT	V_NE - MPH IAS							GROSS WEIGHT
F OAT	8	4	6	8	10	12	14	
0	109	109	105	84	61	--	--	MORE THAN 1,700 LBS
20	109	109	94	72	49	--	--	
40	09	103	81	59	--	--	--	
60	109	91	70	48	--	--	--	
80	109	80	59	--	--	--	--	
100	109	70	48	--	--	--	--	
0	109	109	109	109	98	77	58	1,700 LBS OR LESS
20	109	109	109	109	85	67	48	
40	109	109	109	96	75	57	--	
60	109	109	108	84	66	48	--	
80	109	109	95	74	57	--	--	
100	109	108	84	66	48	--	--	

MAXIMUM V_NE DOORS OFF - 102 MPH IAS

NEVER EXCEED SPEED

V_{NE} KIAS

-20° C
0° C
+20° C
+40° C
MAX ALT.

Pressure Alt. 1,000 Feet

Figure 6-6. Various V_{NE} placards.

Manufacturers also are encouraged to include an optional area titled "Abnormal Procedures," which describes recommended procedures for handling malfunctions that are not considered to be emergencies. This information would most likely be found in larger helicopters.

NORMAL PROCEDURES

The Normal Procedures is the section you will probably use the most. It usually begins with a listing of the airspeeds, which may enhance the safety of normal operations. It is a good idea to memorize the airspeeds that are used for normal flight operations. The next part of the section includes several checklists, which take you through the preflight inspection, before starting procedure, how to start the engine, rotor engagement, ground checks, takeoff, approach, landing, and shutdown. Some manufacturers also include the procedures for practice autorotations. To avoid skipping an important step, you should always use a checklist when one is available. (More information on maneuvers can be found in Chapter 9—Basic Maneuvers, Chapter 10—Advanced Maneuvers, and Chapter 20—Gyroplane Flight Operations.)

PERFORMANCE

The Performance Section contains all the information required by the regulations, and any additional performance information the manufacturer feels may enhance your ability to safely operate the rotorcraft.

These charts, graphs, and tables vary in style but all contain the same basic information. Some examples of the performance information that can be found in most flight manuals include a calibrated versus indicated airspeed conversion graph, hovering ceiling versus gross weight charts, and a height-velocity diagram. [Figure 6-7] For information on how to use the charts, graphs, and tables, refer to Chapter 8—Performance.

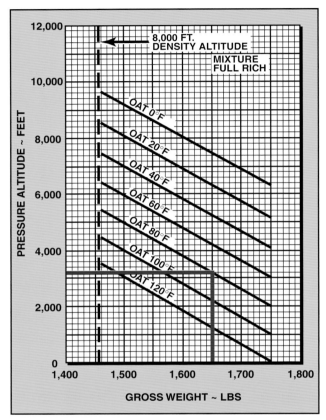

Figure 6-7. One of the performance charts in the Performance Section is the "In Ground Effect Hover Ceiling versus Gross Weight" chart. This chart allows you to determine how much weight you can carry and still operate at a specific pressure altitude, or if you are carrying a specific weight, what is your altitude limitation.

WEIGHT AND BALANCE

The Weight and Balance section should contain all the information required by the FAA that is necessary to calculate weight and balance. To help you correctly compute the proper data, most manufacturers include sample problems. (Weight and balance is further discussed in Chapter 7—Weight and Balance.)

AIRCRAFT AND SYSTEMS DESCRIPTION

The Aircraft and Systems Description section is an excellent place to study and familiarize yourself with all the systems found on your aircraft. The manufactur-

ers should describe the systems in a manner that is understandable to most pilots. For larger, more complex rotorcraft, the manufacturer may assume a higher degree of knowledge. (For more information on rotorcraft systems, refer to Chapter 5—Helicopter Systems and Chapter 18—Gyroplane Systems.)

HANDLING, SERVICING, AND MAINTENANCE

The Handling, Servicing, and Maintenance section describes the maintenance and inspections recommended by the manufacturer, as well as those required by the regulations, and **Airworthiness Directive (AD)** compliance procedures. There are also suggestions on how the pilot/operator can ensure that the work is done properly.

This section also describes preventative maintenance that may be accomplished by certificated pilots, as well as the manufacturer's recommended ground handling procedures, including considerations for hangaring, tie down, and general storage procedures for the rotorcraft.

SUPPLEMENTS

The Supplements Section describes pertinent information necessary to operate optional equipment installed on the rotorcraft that would not be installed on a standard aircraft. Some of this information may be supplied by the aircraft manufacturer, or by the maker of the optional equipment. The information is then inserted into the flight manual at the time the equipment is installed.

SAFETY AND OPERATIONAL TIPS

The Safety and Operational Tips is an optional section that contains a review of information that could enhance the safety of the operation. Some examples of the information that might be covered include: physiological factors, general weather information, fuel conservation procedures, external load warnings, low rotor r.p.m. considerations, and recommendations that if not adhered to could lead to an emergency.

Airworthiness Directive (AD)—A regulatory notice that is sent out by the FAA to the registered owners of aircraft informing them of the discovery of a condition that keeps their aircraft from continuing to meet its conditions for airworthiness. Airworthiness Directives must be complied with within the required time limit, and the fact of compliance, the date of compliance, and the method of compliance must be recorded in the aircraft maintenance records.

It is vital to comply with weight and balance limits established for helicopters. Operating above the maximum weight limitation compromises the structural integrity of the helicopter and adversely affects performance. Balance is also critical because on some fully loaded helicopters, center of gravity deviations as small as three inches can dramatically change a helicopter's handling characteristics. Taking off in a helicopter that is not within the weight and balance limitations is unsafe.

WEIGHT

When determining if your helicopter is within the weight limits, you must consider the weight of the basic helicopter, crew, passengers, cargo, and fuel. Although the effective weight (load factor) varies during maneuvering flight, this chapter primarily considers the weight of the loaded helicopter while at rest.

The following terms are used when computing a helicopter's weight.

BASIC EMPTY WEIGHT—The starting point for weight computations is the basic empty weight, which is the weight of the standard helicopter, optional equipment, unusable fuel, and full operating fluids including full engine oil. Some helicopters might use the term "licensed empty weight," which is nearly the same as basic empty weight, except that it does not include full engine oil, just undrainable oil. If you fly a helicopter that lists a licensed empty weight, be sure to add the weight of the oil to your computations.

USEFUL LOAD—The difference between the gross weight and the basic empty weight is referred to as useful load. It includes the flight crew, usable fuel, drainable oil, if applicable, and payload.

PAYLOAD—The weight of the passengers, cargo, and baggage.

GROSS WEIGHT—The sum of the basic empty weight and useful load.

MAXIMUM GROSS WEIGHT— The maximum weight of the helicopter. Most helicopters have an internal maximum gross weight, which refers to the weight within the helicopter structure and an external maximum gross weight, which refers to the weight of the helicopter with an external load.

WEIGHT LIMITATIONS

Weight limitations are necessary to guarantee the structural integrity of the helicopter, as well as enabling you to predict helicopter performance accurately. Although aircraft manufacturers build in safety factors, you should never intentionally exceed the load limits for which a helicopter is certificated. Operating above a maximum weight could result in structural deformation or failure during flight if you encounter excessive load factors, strong wind gusts, or turbulence. Operating below a minimum weight could adversely affect the handling characteristics of the helicopter. During single-pilot operations in some helicopters, you may have to use a large amount of forward cyclic in order to maintain a hover. By adding ballast to the helicopter, the cyclic will be closer to the center, which gives you a greater range of control motion in every direction. Additional weight also improves autorotational characteristics since the autorotational descent can be established sooner. In addition, operating below minimum weight could prevent you from achieving the desirable rotor r.p.m. during autorotations.

Although a helicopter is certificated for a specified maximum gross weight, it is not safe to take off with this load under all conditions. Anything that adversely affects takeoff, climb, hovering, and landing performance may require off-loading of fuel, passengers, or baggage to some weight less than the published maximum. Factors which can affect performance include high altitude, high temperature, and high humidity conditions, which result in a high density altitude.

DETERMINING EMPTY WEIGHT

A helicopter's weight and balance records contain essential data, including a complete list of all installed optional equipment. Use these records to determine the weight and balance condition of the empty helicopter.

When a helicopter is delivered from the factory, the basic empty weight, empty weight center of gravity (CG), and useful load are recorded on a weight and balance data sheet included in the FAA-Approved Rotorcraft Flight Manual. The basic empty weight can vary even in the same model of helicopter because of differences in installed equipment. If the owner or operator of a helicopter has equipment removed, replaced, or additional equipment installed, these changes must be reflected in the weight and balance records. In addition, major repairs or alterations must be recorded by a certified mechanic. When the revised weight and moment are recorded on a new form, the old record is marked with the word "superseded" and dated with the effective date of the new record. This makes it easy to determine which weight and balance form is the latest version. You must use the latest weight and balance data for computing all loading problems.

BALANCE

Helicopter performance is not only affected by gross weight, but also by the position of that weight. It is essential to load the aircraft within the allowable center-of-gravity range specified in the rotorcraft flight manual's weight and balance limitations.

CENTER OF GRAVITY (CG)

The center of gravity is defined as the theoretical point where all of the aircraft's weight is considered to be concentrated. If a helicopter was suspended by a cable attached to the center-of-gravity point, it would balance like a teeter-totter. For helicopters with a single main rotor, the CG is usually close to the main rotor mast.

Improper balance of a helicopter's load can result in serious control problems. The allowable range in which the CG may fall is called the "CG range." The exact CG location and range are specified in the rotorcraft flight manual for each helicopter. In addition to making a helicopter difficult to control, an out-of-balance loading condition also decreases maneuverability since cyclic control is less effective in the direction opposite to the CG location.

Ideally, you should try to perfectly balance a helicopter so that the fuselage remains horizontal in hovering flight, with no cyclic pitch control needed except for wind correction. Since the fuselage acts as a pendulum suspended from the rotor, changing the center of gravity changes the angle at which the aircraft hangs from the rotor. When the center of gravity is directly under the rotor mast, the helicopter hangs horizontal; if the CG is too far forward of the mast, the helicopter hangs with its nose tilted down; if the CG is too far aft of the mast, the nose tilts up. [Figure 7-1]

CG FORWARD OF FORWARD LIMIT

A forward CG may occur when a heavy pilot and passenger take off without baggage or proper ballast located aft of the rotor mast. This situation becomes worse if the fuel tanks are located aft of the rotor mast because as fuel burns the weight located aft of the rotor mast becomes less.

You can recognize this condition when coming to a hover following a vertical takeoff. The helicopter will have a nose-low attitude, and you will need excessive rearward displacement of the cyclic control to maintain a hover in a no-wind condition. You should not continue flight in this condition, since you could rapidly run out of rearward cyclic control as you consume fuel. You also may find it impossible to decelerate sufficiently to bring the helicopter to a stop. In the event of engine failure and the resulting autorotation, you may not have enough cyclic control to flare properly for the landing.

A forward CG will not be as obvious when hovering into a strong wind, since less rearward cyclic displacement is required than when hovering with no wind. When determining whether a critical balance condition exists, it is essential to consider the wind velocity and its relation to the rearward displacement of the cyclic control.

CG AFT OF AFT LIMIT

Without proper ballast in the cockpit, exceeding the aft CG may occur when:

- A lightweight pilot takes off solo with a full load of fuel located aft of the rotor mast.

- A lightweight pilot takes off with maximum baggage allowed in a baggage compartment located aft of the rotor mast.

- A lightweight pilot takes off with a combination of baggage and substantial fuel where both are aft of the rotor mast.

You can recognize the aft CG condition when coming to a hover following a vertical takeoff. The helicopter will have a tail-low attitude, and you will need exces-

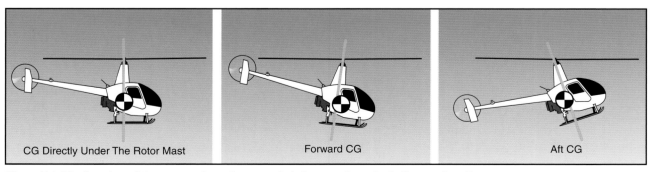

| CG Directly Under The Rotor Mast | Forward CG | Aft CG |

Figure 7-1. The location of the center of gravity strongly influences how the helicopter handles.

sive forward displacement of cyclic control to maintain a hover in a no-wind condition. If there is a wind, you need even greater forward cyclic.

If flight is continued in this condition, you may find it impossible to fly in the upper allowable airspeed range due to inadequate forward cyclic authority to maintain a nose-low attitude. In addition, with an extreme aft CG, gusty or rough air could accelerate the helicopter to a speed faster than that produced with full forward cyclic control. In this case, dissymmetry of lift and blade flapping could cause the rotor disc to tilt aft. With full forward cyclic control already applied, you might not be able to lower the rotor disc, resulting in possible loss of control, or the rotor blades striking the tailboom.

LATERAL BALANCE
For most helicopters, it is usually not necessary to determine the lateral CG for normal flight instruction and passenger flights. This is because helicopter cabins are relatively narrow and most optional equipment is located near the center line. However, some helicopter manuals specify the seat from which you must conduct solo flight. In addition, if there is an unusual situation, such as a heavy pilot and a full load of fuel on one side of the helicopter, which could affect the lateral CG, its position should be checked against the CG envelope. If carrying external loads in a position that requires large lateral cyclic control displacement to maintain level flight, fore and aft cyclic effectiveness could be dramatically limited.

WEIGHT AND BALANCE CALCULATIONS
When determining whether your helicopter is properly loaded, you must answer two questions:

1. Is the gross weight less than or equal to the maximum allowable gross weight?

2. Is the center of gravity within the allowable CG range, and will it stay within the allowable range as fuel is burned off?

To answer the first question, just add the weight of the items comprising the useful load (pilot, passengers, fuel, oil, if applicable, cargo, and baggage) to the basic empty weight of the helicopter. Check that the total weight does not exceed the maximum allowable gross weight.

To answer the second question, you need to use CG or moment information from loading charts, tables, or graphs in the rotorcraft flight manual. Then using one of the methods described below, calculate the loaded moment and/or loaded CG and verify that it falls within the allowable CG range shown in the rotorcraft flight manual.

It is important to note that any weight and balance computation is only as accurate as the information provided. Therefore, you should ask passengers what they weigh

and add a few pounds to cover the additional weight of clothing, especially during the winter months. The baggage weight should be determined by the use of a scale, if practical. If a scale is not available, be conservative and overestimate the weight. Figure 7-2 indicates the standard weights for specific operating fluids.

Aviation Gasoline (AVGAS)	.6 lbs. / gal.
Jet Fuel (JP-4)	6.5 lbs. / gal.
Jet Fuel (JP-5)	6.8 lbs. / gal.
Reciprocating Engine Oil	7.5 lbs. / gal.*
Turbine Engine Oil . . Varies between 7.5 and 8.5 lbs. / gal.*
Water	.8.35 lbs. / gal.

* Oil weight is given in pounds per gallon while oil capacity is usually given in quarts; therefore, you must convert the amount of oil to gallons before calculating its weight.

Figure 7-2. When making weight and balance computations, always use actual weights if they are available, especially if the helicopter is loaded near the weight and balance limits.

The following terms are used when computing a helicopter's balance.

REFERENCE DATUM—Balance is determined by the location of the CG, which is usually described as a given number of inches from the reference datum. The horizontal reference datum is an imaginary vertical plane or point, arbitrarily fixed somewhere along the longitudinal axis of the helicopter, from which all horizontal distances are measured for weight and balance purposes. There is no fixed rule for its location. It may be located at the rotor mast, the nose of the helicopter, or even at a point in space ahead of the helicopter. [Figure 7-3]

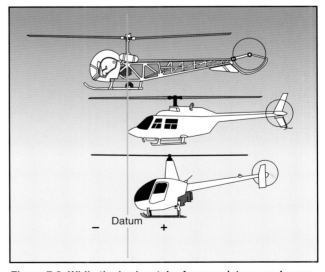

Figure 7-3. While the horizontal reference datum can be anywhere the manufacturer chooses, most small training helicopters have the horizontal reference datum 100 inches forward of the main rotor shaft centerline. This is to keep all the computed values positive.

The lateral reference datum, is usually located at the center of the helicopter. The location of the reference datums is established by the manufacturer and is defined in the rotorcraft flight manual. [Figure 7-4]

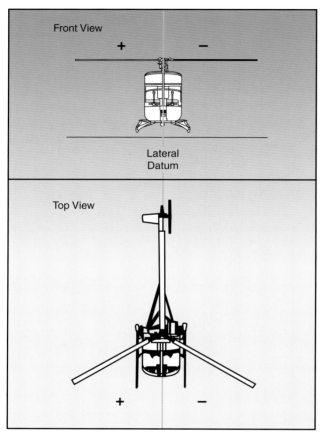

Figure 7-4. The lateral reference datum is located longitudinally through the center of the helicopter; therefore, there are positive and negative values.

ARM—The horizontal distance from the datum to any component of the helicopter or to any object located within the helicopter is called the arm. Another term that can be used interchangeably with arm is station. If the component or object is located to the rear of the datum, it is measured as a positive number and usually is referred to as inches aft of the datum. Conversely, if the component or object is located forward of the datum, it is indicated as a negative number and is usually referred to as inches forward of the datum.

MOMENT—If the weight of an object is multiplied by its arm, the result is known as its moment. You may think of moment as a force that results from an object's weight acting at a distance. Moment is also referred to as the tendency of an object to rotate or pivot about a point. The farther an object is from a pivotal point, the greater its force.

CENTER OF GRAVITY COMPUTATION—By totaling the weights and moments of all components and objects carried, you can determine the point where a loaded helicopter would balance. This point is known as the center of gravity.

WEIGHT AND BALANCE METHODS
Since weight and balance is so critical to the safe operation of a helicopter, it is important to know how to check this condition for each loading arrangement. Most helicopter manufacturers use one of two methods, or a combination of the methods, to check weight and balance conditions.

COMPUTATIONAL METHOD
With the computational method, you use simple mathematics to solve weight and balance problems. The first step is to look up the basic empty weight and total moment for the particular helicopter you fly. If the center of gravity is given, it should also be noted. The empty weight CG can be considered the arm of the empty helicopter. This should be the first item recorded on the weight and balance form. [Figure 7-5]

	Weight (pounds)	Arm (inches)	Moment (lb/inches)
Basic Empty Weight	1,700	116.5	198,050
Oil	12	179.0	2,148
Pilot	190	65.0	12,350
Forward Passenger	170	65.0	11,050
Passengers Aft	510	104	53,040
Baggage	40	148	5,920
Fuel	553	120	66,360
Total	3,175		348,918
CG		109.9	
Max Gross Weight = 3,200 lbs. CG Range 106.0 – 114.2 in.			

Figure 7-5. In this example, the helicopter's weight of 1,700 pounds is recorded in the first column, its CG or arm of 116.5 inches in the second, and its moment of 198,050 pound-inches in the last. Notice that the weight of the helicopter, multiplied by its CG, equals its moment.

Next, the weights of the oil, if required, pilot, passengers, baggage, and fuel are recorded. Use care in recording the weight of each passenger and baggage. Recording each weight in its proper location is extremely important to the accurate calculation of a CG. Once you have recorded all of the weights, add them together to determine the total weight of the loaded helicopter.

Now, check to see that the total weight does not exceed the maximum allowable weight under existing conditions. In this case, the total weight of the helicopter is under the maximum gross weight of 3,200 pounds.

Once you are satisfied that the total weight is within prescribed limits, multiply each individual weight by its associated arm to determine its moment. Then, add the moments together to arrive at the total moment for the helicopter. Your final computation is to find the center of gravity of the loaded helicopter by dividing the total moment by the total weight.

After determining the helicopter's weight and center of gravity location, you need to determine if the CG is within acceptable limits. In this example, the allowable range is between 106.0 inches and 114.2 inches. Therefore, the CG location is within the acceptable range. If the CG falls outside the acceptable limits, you will have to adjust the loading of the helicopter.

LOADING CHART METHOD

You can determine if a helicopter is within weight and CG limits using a loading chart similar to the one in figure 7-6. To use this chart, first subtotal the empty weight, pilot, and passengers. This is the weight at which you enter the chart on the left. The next step is to follow the upsloping lines for baggage and then for fuel to arrive at your final weight and CG. Any value on or inside the envelope is within the range.

SAMPLE PROBLEM 1

Determine if the gross weight and center of gravity are within allowable limits under the following loading conditions for a helicopter based on the loading chart in figure 7-6.

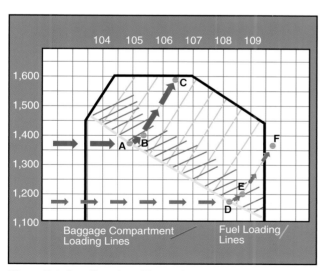

Figure 7-6. Loading chart illustrating the solution to sample problems 1 and 2.

To use the loading chart for the helicopter in this example, you must add up the items in a certain order. The maximum allowable gross weight is 1,600 pounds.

ITEM	POUNDS
Basic empty weight	1,040
Pilot	135
Passenger	200
Subtotal	1,375 (point A)
Baggage compartment load	25
Subtotal	1,400 (point B)
Fuel load (30 gallons)	180
Total weight	1,580 (point C)

1. Follow the green arrows in figure 7-6. Enter the graph on the left side at 1,375 lb., the subtotal of the empty weight and the passenger weight. Move right to the yellow line. (point A)

2. Move up and to the right, parallel to the baggage compartment loading lines to 1,400 lb. (Point B)

3. Continue up and to the right, this time parallel to the fuel loading lines, to the total weight of 1,580 lb. (Point C).

Point C is within allowable weight and CG limits.

SAMPLE PROBLEM 2

Assume that the pilot in sample problem 1 discharges the passenger after using only 20 pounds of fuel.

ITEM	POUNDS
Basic empty weight	1,040
Pilot	135
Subtotal	1,175 (point D)
Baggage compartment load	25
Subtotal	1,200 (point E)
Fuel load	160
Total weight	1,360 (point F)

Follow the blue arrows in figure 7-6, starting at 1,175 lb. on the left side of the graph, then to point D, E, and F. Although the total weight of the helicopter is well below the maximum allowable gross weight, point F falls outside the aft allowable CG limit.

As you can see, it is important to reevaluate the balance in a helicopter whenever you change the loading. Unlike most airplanes, where discharging a passenger is unlikely to adversely affect the CG, off-loading a passenger from a helicopter could make the aircraft unsafe to fly. Another difference between helicopter and airplane loading is that most small airplanes carry fuel in the wings very near the center of gravity. Burning off fuel has little effect on the loaded CG. However, helicopter fuel tanks are often significantly behind the center of gravity. Consuming fuel from a tank aft of the rotor mast causes the loaded helicopter CG to move forward. As standard practice, you should compute the weight and balance with zero fuel to verify that your helicopter remains within the acceptable limits as fuel is used.

SAMPLE PROBLEM 3

The loading chart used in the sample problems 1 and 2 is designed to graphically calculate the loaded center of gravity and show whether it is within limits, all on a single chart. Another type of loading chart calculates moments for each station. You must then add up these moments and consult another graph to determine whether the total is within limits. Although this method has more steps, the charts are sometimes easier to use.

To begin, record the basic empty weight of the helicopter, along with its total moment. Remember to use the actual weight and moment of the helicopter you are flying. Next, record the weights of the pilot, passengers, fuel, and baggage on a weight and balance worksheet. Then, determine the total weight of the helicopter. Once you have determined the weight to be within prescribed limits, compute the moment for each weight and for the loaded helicopter. Do this with a loading graph provided by the manufacturer. Use figure 7-7 to determine the moments for a pilot and passenger weighing 340 pounds and for 211 pounds of fuel.

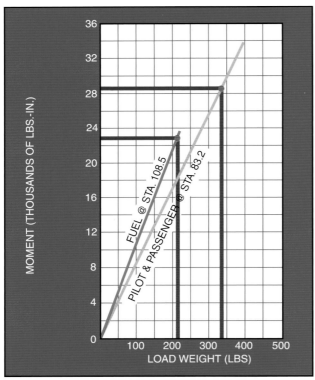

Figure 7-7. Moments for fuel, pilot, and passenger.

Start at the bottom scale labeled LOAD WEIGHT. Draw a line from 211 pounds up to the line labeled "FUEL @ STA108.5." Draw your line to the left to intersect the MOMENT scale and read the fuel moment (22.9 thousand lb.-inches). Do the same for the pilot/passenger moment. Draw a line from a weight of 340 pounds up to the line labeled "PILOT & PASSENGER

@STA. 83.2." Go left and read the pilot/passenger moment (28.3 thousand lb.-inches).

Reduction factors are often used to reduce the size of large numbers to manageable levels. In figure 7-7, the scale on the loading graph gives you moments in thousands of pound-inches. In most cases, when using this type of chart, you need not be concerned with reduction factors because the CG/moment envelope chart normally uses the same reduction factor. [Figure 7-8]

	Weight (lbs.)	Moment (lb.-ins. /1,000)
1. Basic Empty Weight..............	1,102	110.8
2. Pilot and Front Passenger........	340	28.3
3. Fuel............................	211	22.9
5. Baggage................................		
TOTALS	1,653	162.0

Figure 7-8. CG/Moment Chart.

After recording the basic empty weight and moment of the helicopter, and the weight and moment for each item, total and record all weights and moments. Next, plot the calculated takeoff weight and moment on the sample moment envelope graph. Based on a weight of 1,653 pounds and a moment/1,000 of 162 pound-inches, the helicopter is within the prescribed CG limits.

COMBINATION METHOD

The combination method usually uses the computation method to determine the moments and center of gravity. Then, these figures are plotted on a graph to determine if they intersect within the acceptable envelope. Figure 7-9 illustrates that with a total weight of 2,399 pounds and a total moment of 225,022 pound-

	Weight (pounds)	Longitudinal	
		Arm (inches)	Moment (lb/inches)
Basic Empty Weight	1,400	107.75	150,850
Pilot	170	49.5	8,415
Fwd Passenger	250	49.5	12,375
Right Fwd Baggage		44	0
Left Fwd Baggage		44	0
Right Aft Passenger		79.5	0
Left Aft Passenger	185	79.5	14,708
Right Aft Baggage	50	79.5	3,975
Left Aft Baggage	50	79.5	3,975
Totals with Zero Fuel	2,105		194,298
Main Fuel Tank	184	106	19,504
Aux Fuel Tank	110	102	11,220
Totals with Fuel	2,399		225,022
CG		93.8	

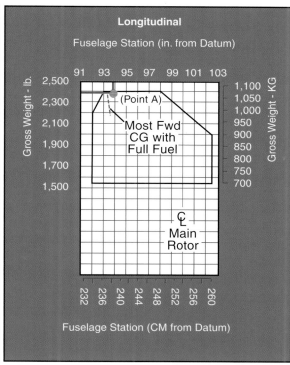

Figure 7-9. Use the longitudinal CG envelope along with the computed CGs to determine if the helicopter is loaded properly.

inches, the CG is 93.8. Plotting this CG against the weight indicates that the helicopter is loaded within the longitudinal limits (point A).

CALCULATING LATERAL CG

Some helicopter manufacturers require that you also determine the lateral CG limits. These calculations are similar to longitudinal calculations. However, since the lateral CG datum line is almost always defined as the center of the helicopter, you are likely to encounter negative CGs and moments in your calculations. Negative values are located on the left side while positive stations are located on the right.

Refer to figure 7-10. When computing moment for the pilot, 170 pounds is multiplied by the arm of 12.2 inches resulting in a moment of 2,074 pound-inches. As with any weight placed right of the aircraft centerline, the moment is expressed as a positive value. The forward passenger sits left of the aircraft centerline. To compute this moment, multiply 250 pounds by –10.4 inches. The result is in a moment of –2,600 pound-inches. Once the aircraft is completely loaded, the weights and moments are totaled and the CG is computed. Since more weight is located left of the aircraft centerline, the resulting total moment is –3,837 pound-inches. To calculate CG, divide –3,837 pound-inches by the total weight of 2,399 pounds. The result is –1.6 inches, or a CG that is 1.6 inches left of the aircraft centerline.

	Weight (pounds)	Lateral	
		Arm (inches)	Moment (lb/inches)
Basic Empty Weight	1,400	0	0
Pilot	170	12.2	2,074
Fwd Passenger	250	–10.4	–2,600
Right Fwd Baggage		11.5	0
Left Fwd Baggage		–11.5	0
Right Aft Passenger		12.2	0
Left Aft Passenger	185	–12.2	–2,257
Right Aft Baggage	50	12.2	610
Left Aft Baggage	50	–12.2	–610
Totals with Zero Fuel	2,105		–2,783
Main Fuel Tank	184	–13.5	–2,484
Aux Fuel Tank	110	13	1,430
Totals with Fuel	2,399		–3,837
CG		–1.6	

Figure 7-10. Computed Lateral CG.

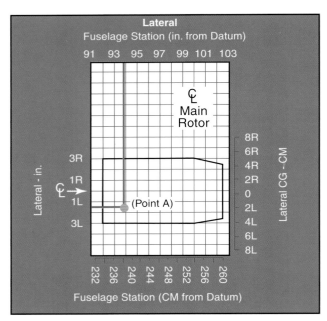

Figure 7-11. Use the lateral CG envelope to determine if the helicopter is properly loaded.

Lateral CG is often plotted against the longitudinal CG. [Figure 7-11] In this case, −1.6 is plotted against 93.8, which was the longitudinal CG determined in the previous problem. The intersection of the two lines falls well within the lateral CG envelope.

Your ability to predict the performance of a helicopter is extremely important. It allows you to determine how much weight the helicopter can carry before takeoff, if your helicopter can safely hover at a specific altitude and temperature, how far it will take to climb above obstacles, and what your maximum climb rate will be.

FACTORS AFFECTING PERFORMANCE

A helicopter's performance is dependent on the power output of the engine and the lift production of the rotors, whether it is the main rotor(s) or tail rotor. Any factor that affects engine and rotor efficiency affects performance. The three major factors that affect performance are density altitude, weight, and wind.

DENSITY ALTITUDE

The density of the air directly affects the performance of the helicopter. As the density of the air increases, engine power output, rotor efficiency, and aerodynamic lift all increase. **Density altitude** is the altitude above mean sea level at which a given atmospheric density occurs in the **standard atmosphere**. It can also be interpreted as **pressure altitude** corrected for nonstandard temperature differences.

Pressure altitude is displayed as the height above a standard datum plane, which, in this case, is a theoretical plane where air pressure is equal to 29.92 in. Hg. Pressure altitude is the indicated height value on the altimeter when the altimeter setting is adjusted to 29.92 in. Hg. Pressure altitude, as opposed to **true altitude**, is an important value for calculating performance as it more accurately represents the air content at a particular level. The difference between true altitude

and pressure altitude must be clearly understood. True altitude means the vertical height above mean sea level and is displayed on the altimeter when the altimeter is correctly adjusted to the local setting.

For example, if the local altimeter setting is 30.12 in. Hg., and the altimeter is adjusted to this value, the altimeter indicates exact height above sea level. However, this does not reflect conditions found at this height under standard conditions. Since the altimeter setting is more than 29.92 in. Hg., the air in this example has a higher pressure, and is more compressed, indicative of the air found at a lower altitude. Therefore, the pressure altitude is lower than the actual height above mean sea level.

To calculate pressure altitude without the use of an altimeter, remember that the pressure decreases approximately 1 inch of mercury for every 1,000-foot increase in altitude. For example, if the current local altimeter setting at a 4,000-foot elevation is 30.42, the pressure altitude would be 3,500 feet. (30.42 – 29.92 = .50 in. Hg. 3 1,000 feet = 500 feet. Subtracting 500 feet from 4,000 equals 3,500 feet).

The four factors that most affect density altitude are: atmospheric pressure, altitude, temperature, and the moisture content of the air.

ATMOSPHERIC PRESSURE

Due to changing weather conditions, atmospheric pressure at a given location changes from day to day. If the pressure is lower, the air is less dense. This means a higher density altitude and less helicopter performance.

Density Altitude—Pressure altitude corrected for nonstandard temperature variations. Performance charts for many older aircraft are based on this value.

Standard Atmosphere—At sea level, the standard atmosphere consists of a barometric pressure of 29.92 inches of mercury (in. Hg.) or 1013.2 millibars, and a temperature of 15°C (59°F). Pressure and temperature normally decrease as altitude increases. The standard lapse rate in the lower atmosphere for each 1,000 feet of altitude is approximately 1 in. Hg. and 2°C (3.5°F). For example, the standard pressure and temperature at 3,000 feet mean sea level (MSL) is 26.92 in. Hg. (29.92 – 3) and 9°C (15°C – 6°C).

Pressure Altitude—The height above the standard pressure level of 29.92 in. Hg. It is obtained by setting 29.92 in the barometric pressure window and reading the altimeter.

True Altitude—The actual height of an object above mean sea level.

ALTITUDE

As altitude increases, the air becomes thinner or less dense. This is because the atmospheric pressure acting on a given volume of air is less, allowing the air molecules to move further apart. Dense air contains more air molecules spaced closely together, while thin air contains less air molecules because they are spaced further apart. As altitude increases, density altitude increases.

TEMPERATURE

Temperature changes have a large affect on density altitude. As warm air expands, the air molecules move further apart, creating less dense air. Since cool air contracts, the air molecules move closer together, creating denser air. High temperatures cause even low elevations to have high density altitudes.

MOISTURE (HUMIDITY)

The water content of the air also changes air density because water vapor weighs less than dry air. Therefore, as the water content of the air increases, the air becomes less dense, increasing density altitude and decreasing performance.

Humidity, also called "relative humidity," refers to the amount of water vapor contained in the atmosphere, and is expressed as a percentage of the maximum amount of water vapor the air can hold. This amount varies with temperature; warm air can hold more water vapor, while colder air can hold less. Perfectly dry air that contains no water vapor has a relative humidity of 0 percent, while saturated air that cannot hold any more water vapor, has a relative humidity of 100 percent.

Humidity alone is usually not considered an important factor in calculating density altitude and helicopter performance; however, it does contribute. There are no rules-of-thumb or charts used to compute the effects of humidity on density altitude, so you need to take this into consideration by expecting a decrease in hovering and takeoff performance in high humidity conditions.

HIGH AND LOW DENSITY ALTITUDE CONDITIONS

You need to thoroughly understand the terms "high density altitude" and "low density altitude." In general, high density altitude refers to thin air, while low density altitude refers to dense air. Those conditions that result in a high density altitude (thin air) are high elevations, low atmospheric pressure, high temperatures, high humidity, or some combination thereof. Lower elevations, high atmospheric pressure, low temperatures, and low humidity are more indicative of low density altitude (dense air). However, high density altitudes may be present at lower elevations on hot days, so it is important to calculate the density altitude and determine performance before a flight.

One of the ways you can determine density altitude is through the use of charts designed for that purpose. [Figure 8-1]. For example, assume you are planning to depart an airport where the field elevation is 1,165 feet MSL, the altimeter setting is 30.10, and the temperature is 70°F. What is the density altitude? First, correct for nonstandard pressure (30.10) by referring to the right side of the chart, and subtracting 165 feet from the field elevation. The result is a pressure altitude of 1,000 feet. Then, enter the chart at the bottom, just above the temperature of 70°F (21°C). Proceed up the chart vertically until you intercept the diagonal 1,000-foot pressure altitude line, then move horizontally to the left and read the density altitude of approximately 2,000 feet. This means your helicopter will perform as if it were at 2,000 feet MSL on a standard day.

Most performance charts do not require you to compute density altitude. Instead, the computation is built into the performance chart itself. All you have to do is enter the chart with the correct pressure altitude and the temperature.

WEIGHT

Lift is the force that opposes weight. As weight increases, the power required to produce the lift needed to compensate for the added weight must also increase. Most performance charts include weight as one of the variables. By reducing the weight of the helicopter, you may find that you are able to safely take off or land at a location that otherwise would be impossible. However, if you are ever in doubt about whether you can safely perform a takeoff or landing, you should delay your takeoff until more favorable density altitude conditions exist. If airborne, try to land at a location that has more favorable conditions, or one where you can make a landing that does not require a hover.

In addition, at higher gross weights, the increased power required to hover produces more torque, which means more antitorque thrust is required. In some helicopters, during high altitude operations, the maximum antitorque produced by the tail rotor during a hover may not be sufficient to overcome torque even if the gross weight is within limits.

WINDS

Wind direction and velocity also affect hovering, takeoff, and climb performance. Translational lift occurs anytime there is relative airflow over the rotor disc. This occurs whether the relative airflow is caused by helicopter movement or by the wind. As wind speed increases, translational lift increases, resulting in less power required to hover.

The wind direction is also an important consideration. Headwinds are the most desirable as they contribute to the most increase in performance. Strong crosswinds

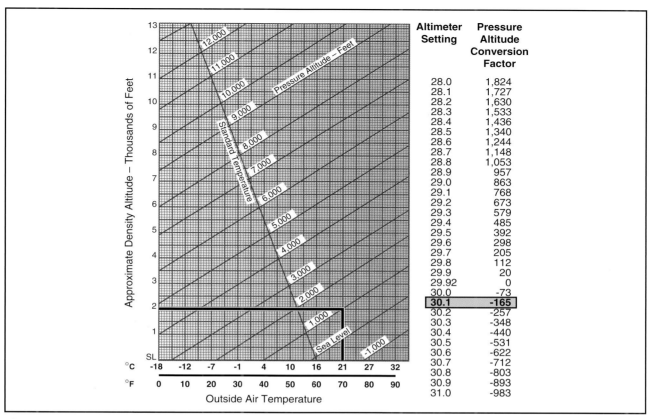

Figure 8-1. Density Altitude Chart.

and tailwinds may require the use of more tail rotor thrust to maintain directional control. This increased tail rotor thrust absorbs power from the engine, which means there is less power available to the main rotor for the production of lift. Some helicopters even have a critical wind azimuth or maximum safe relative wind chart. Operating the helicopter beyond these limits could cause loss of tail rotor effectiveness.

Takeoff and climb performance is greatly affected by wind. When taking off into a headwind, effective translational lift is achieved earlier, resulting in more lift and a steeper climb angle. When taking off with a tailwind, more distance is required to accelerate through translation lift.

PERFORMANCE CHARTS

In developing performance charts, aircraft manufacturers make certain assumptions about the condition of the helicopter and the ability of the pilot. It is assumed that the helicopter is in good operating condition and the engine is developing its rated power. The pilot is assumed to be following normal operating procedures and to have average flying abilities. Average means a pilot capable of doing each of the required tasks correctly and at the appropriate times.

Using these assumptions, the manufacturer develops performance data for the helicopter based on

actual flight tests. However, they do not test the helicopter under each and every condition shown on a performance chart. Instead, they evaluate specific data and mathematically derive the remaining data.

HOVERING PERFORMANCE

Helicopter performance revolves around whether or not the helicopter can be hovered. More power is required during the hover than in any other flight regime. Obstructions aside, if a hover can be maintained, a takeoff can be made, especially with the additional benefit of translational lift. Hover charts are provided for **in ground effect (IGE) hover** and **out of ground effect (OGE) hover** under various conditions of gross weight, altitude, temperature, and power. The "in ground effect" hover ceiling is usually higher than the "out of ground effect" hover ceiling because of the added lift benefit produced by ground effect.

In Ground Effect (IGE) Hover—Hovering close to the surface (usually less than one rotor diameter above the surface) under the influence of ground effect.

Out of Ground Effect (OGE) Hover—Hovering greater than one rotor diameter distance above the surface. Because induced drag is greater while hovering out of ground effect, it takes more power to achieve a hover. See Chapter 3—Aerodynamics of Flight for more details on IGE and OGE hover.

As density altitude increases, more power is required to hover. At some point, the power required is equal to the power available. This establishes the hovering ceiling under the existing conditions. Any adjustment to the gross weight by varying fuel, payload, or both, affects the hovering ceiling. The heavier the gross weight, the lower the hovering ceiling. As gross weight is decreased, the hover ceiling increases.

SAMPLE PROBLEM 1

You are to fly a photographer to a remote location to take pictures of the local wildlife. Using figure 8-2, can you safely hover in ground effect at your departure point with the following conditions?

Pressure Altitude...................................8,000 feet
Temperature..+15°C
Takeoff Gross Weight.....................1,250 pounds
R.P.M..104%

First enter the chart at 8,000 feet pressure altitude (point A), then move right until reaching a point midway between the +10°C and +20°C lines (point B). From that point, proceed down to find the maximum gross weight where a 2 foot hover can be achieved. In this case, it is approximately 1,280 pounds (point C).

Since the gross weight of your helicopter is less than this, you can safely hover with these conditions.

SAMPLE PROBLEM 2

Once you reach the remote location in the previous problem, you will need to hover out of ground effect for some of the pictures. The pressure altitude at the remote site is 9,000 feet, and you will use 50 pounds of fuel getting there. (The new gross weight is now 1,200 pounds.) The temperature will remain at +15°C. Using figure 8-3, can you accomplish the mission?

Enter the chart at 9,000 feet (point A) and proceed to point B (+15°C). From there determine that the maximum gross weight to hover out of ground effect is approximately 1,130 pounds (point C). Since your gross weight is higher than this value, you will not be able to hover with these conditions. To accomplish the mission, you will have to remove approximately 70 pounds before you begin the flight.

These two sample problems emphasize the importance of determining the gross weight and hover ceiling throughout

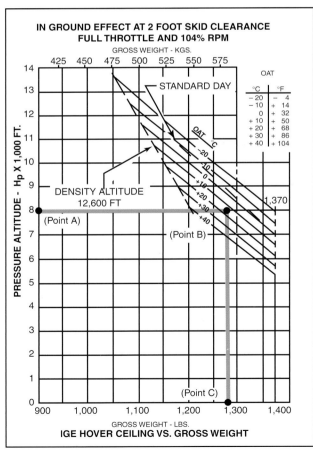

Figure 8-2. In Ground Effect Hover Ceiling versus Gross Weight Chart.

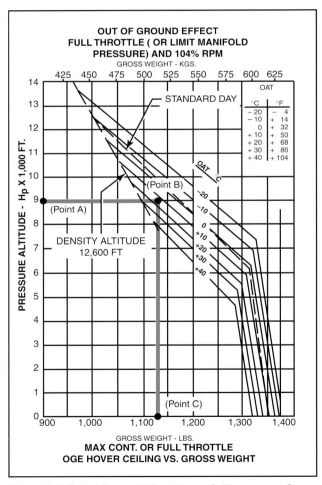

Figure 8-3. Out of Ground Effect Hover Ceiling versus Gross Weight Chart.

the entire flight operation. Being able to hover at the take-off location with a certain gross weight does not ensure the same performance at the landing point. If the destination point is at a higher density altitude because of higher elevation, temperature, and/or relative humidity, more power is required to hover. You should be able to predict whether hovering power will be available at the destination by knowing the temperature and wind conditions, using the performance charts in the helicopter flight manual, and making certain power checks during hover and in flight prior to commencing the approach and landing.

TAKEOFF PERFORMANCE

If takeoff charts are included in the rotorcraft flight manual, they usually indicate the distance it takes to clear a 50-foot obstacle based on various conditions of weight, pressure altitude, and temperature. In addition, the values computed in the takeoff charts usually assume that the flight profile is per the applicable height-velocity diagram.

SAMPLE PROBLEM 3

In this example, determine the distance to clear a 50-foot obstacle with the following conditions:

Pressure Altitude..................................5,000 feet
Takeoff Gross Weight....................2,850 pounds
Temperature ....95°F

Using figure 8-4, locate 2,850 pounds in the first column. Since the pressure altitude of 5,000 feet is not one of the choices in column two, you have to interpolate between the values from the 4,000- and 6,000-foot lines. Follow each of these rows out to the column

headed by 95°F. The values are 1,102 feet and 1,538 feet. Since 5,000 is halfway between 4,000 and 6,000, the interpolated value should be halfway between these two values or 1,320 feet ([1,102 + 1,538] 4 2 = 1,320).

CLIMB PERFORMANCE

Most of the factors affecting hover and takeoff performance also affect climb performance. In addition, turbulent air, pilot techniques, and overall condition of the helicopter can cause climb performance to vary.

A helicopter flown at the "best rate-of-climb" speed will obtain the greatest gain in altitude over a given period of time. This speed is normally used during the climb after all obstacles have been cleared and is usually maintained until reaching cruise altitude. Rate of climb must not be confused with angle of climb. Angle of climb is a function of altitude gained over a given distance. The best rate-of-climb speed results in the highest climb rate, but not the steepest climb angle and may not be sufficient to clear obstructions. The "best angle-of-climb" speed depends upon the power available. If there is a surplus of power available, the helicopter can climb vertically, so the best angle-of-climb speed is zero.

Wind direction and speed have an effect on climb performance, but it is often misunderstood. Airspeed is the speed at which the helicopter is moving through the atmosphere and is unaffected by wind. Atmospheric wind affects only the groundspeed, or speed at which the helicopter is moving over the earth's surface. Thus, the only climb performance

TAKE-OFF DISTANCE (FEET TO CLEAR 50 FOOT OBSTACLE)					
Gross Weight Pounds	Pressure Altitude Feet	At −13°F −25°C	At 23°F −5°C	At 59°F 15°C	At 95°F 35°C
2,150	SL	373	401	430	458
	2,000	400	434	461	491
	4,000	428	462	494	527
	6,000	461	510	585	677
	8,000	567	674	779	896
2,500	SL	531	569	613	652
	2,000	568	614	660	701
	4,000	611	660	709	759
	6,000	654	727	848	986
	8,000	811	975	1,144	1,355
2,850	SL	743	806	864	929
	2,000	770	876	929	1,011
	4,000	861	940	1,017	1,102 / 1,320
	6,000	939	1,064	1,255	1,538
	8,000	1,201	1,527	–	–

Figure 8-4. Takeoff Distance Chart.

affected by atmospheric wind is the angle of climb and not the rate of climb.

SAMPLE PROBLEM 4

Determine the best rate of climb using figure 8-5. Use the following conditions:

Pressure Altitude ....12,000 feet
Outside Air Temperature ....+10°C
Gross Weight ....3,000 pounds
Power ....Takeoff Power
Anti-ice ....ON
Indicated Airspeed ....52 knots

With this chart, first locate the temperature of +10°C (point A). Then proceed up the chart to the 12,000-foot pressure altitude line (point B). From there, move horizontally to the right until you intersect the 3,000-foot line (point C). With this performance chart, you must now determine the rate of climb with anti-ice off and then subtract the rate of climb change with it on. From point C, go to the bottom of the chart and find that the maximum rate of climb with anti-ice off is approximately 890 feet per minute. Then, go back to point C and up to the anti-ice-on line (point D). Proceed horizontally to the right and read approximately 240 feet per minute change (point E). Now subtract 240 from 890 to get a maximum rate of climb, with anti-ice on, of 650 feet per minute.

Other rate-of-climb charts use density altitude as a starting point. [Figure 8-6] While it cleans up the chart somewhat, you must first determine density altitude. Notice also that this chart requires a change in the indicated airspeed with a change in altitude.

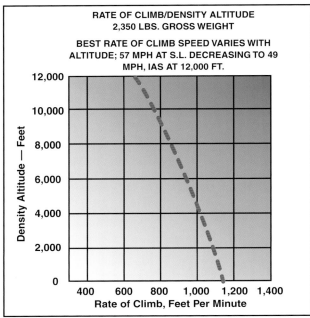

Figure 8-6. This chart uses density altitude in determining maximum rate of climb.

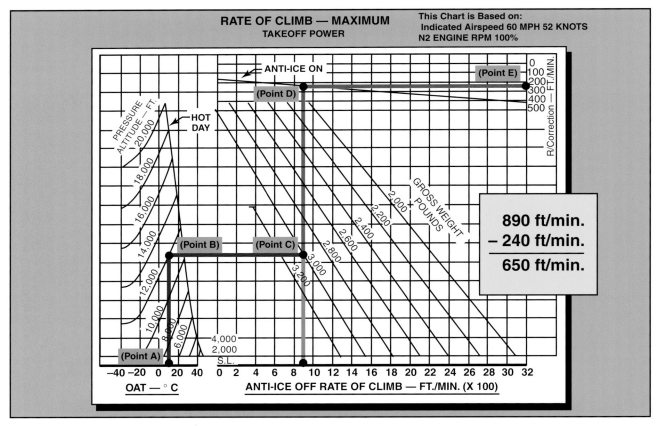

Figure 8-5. Maximum Rate-of-Climb Chart.

Basic Flight Maneuvers

From the previous chapters, it should be apparent that no two helicopters perform the same way. Even when flying the same model of helicopter, wind, temperature, humidity, weight, and equipment make it difficult to predict just how the helicopter will perform. Therefore, this chapter presents the basic flight maneuvers in a way that would apply to a majority of the helicopters. In most cases, the techniques described apply to small training helicopters with:

- A single, main rotor rotating in a counterclockwise direction (looking downward on the rotor).

- An antitorque system.

Where a technique differs, it will be noted. For example, a power increase on a helicopter with a clockwise rotor system requires right antitorque pedal pressure instead of left pedal pressure. In many cases, the terminology "apply proper pedal pressure" is used to indicate both types of rotor systems. However, when discussing throttle coordination to maintain proper r.p.m., there will be no differentiation between those helicopters with a governor and those without. In a sense, the governor is doing the work for you. In addition, instead of using the terms collective pitch control and the cyclic pitch control throughout the chapter, these controls are referred to as just collective and cyclic.

Because helicopter performance varies with different weather conditions and aircraft loading, specific nose attitudes and power settings will not be discussed. In addition, this chapter does not detail each and every attitude of a helicopter in the various flight maneuvers, nor each and every move you must make in order to perform a given maneuver.

When a maneuver is presented, there will be a brief description, followed by the technique to accomplish the maneuver. In most cases, there is a list of common errors at the end of the discussion.

PREFLIGHT

Before any flight, you must ensure the helicopter is airworthy by inspecting it according to the rotorcraft flight manual, pilot's operating handbook, or other information supplied either by the operator or the manufacturer. Remember that as pilot in command, it is your responsibility to ensure the aircraft is in an airworthy condition.

In preparation for flight, the use of a checklist is important so that no item is overlooked. Follow the manufacturer's suggested outline for both the inside and outside inspection. This ensures that all the items the manufacturer feels are important are checked. Obviously, if there are other items you feel might need attention, inspect them as well.

MINIMUM EQUIPMENT LISTS (MELS) AND OPERATIONS WITH INOPERATIVE EQUIPMENT

The Code of Federal Regulations (CFRs) requires that all aircraft instruments and installed equipment be operative prior to each departure. However, when the FAA adopted the **minimum equipment list (MEL)** concept for 14 CFR part 91 operations, flights were allowed with inoperative items, as long as the inoperative items were determined to be nonessential for safe flight. At the same time, it allowed part 91 operators, without an MEL, to defer repairs on nonessential equipment within the guidelines of part 91.

There are two primary methods of deferring maintenance on rotorcraft operating under part 91. They are the deferral provision of 14 CFR part 91, section 91.213(d) and an FAA-approved MEL.

The deferral provision of section 91.213(d) is widely used by most pilot/operators. Its popularity is due to simplicity and minimal paperwork. When inoperative equipment is found during preflight or prior to departure, the decision should be to cancel the flight, obtain maintenance prior to flight, or to defer the item or equipment.

Maintenance deferrals are not used for in-flight discrepancies. The manufacturer's RFM/POH procedures are to be used in those situations. The discussion that

Minimum Equipment List (MEL)—An inventory of instruments and equipment that may legally be inoperative, with the specific conditions under which an aircraft may be flown with such items inoperative.

follows assumes that the pilot wishes to defer maintenance that would ordinarily be required prior to flight.

Using the deferral provision of section 91.213(d), the pilot determines whether the inoperative equipment is required by type design, the CFRs, or ADs. If the inoperative item is not required, and the helicopter can be safely operated without it, the deferral may be made. The inoperative item shall be deactivated or removed and an INOPERATIVE placard placed near the appropriate switch, control, or indicator. If deactivation or removal involves maintenance (removal always will), it must be accomplished by certificated maintenance personnel.

For example, if the position lights (installed equipment) were discovered to be inoperative prior to a daytime flight, the pilot would follow the requirements of section 91.213(d).

The deactivation may be a process as simple as the pilot positioning a circuit breaker to the OFF position, or as complex as rendering instruments or equipment totally inoperable. Complex maintenance tasks require a certificated and appropriately rated maintenance person to perform the deactivation. In all cases, the item or equipment must be placarded INOPERATIVE.

All rotorcraft operated under part 91 are eligible to use the maintenance deferral provisions of section 91.213(d). However, once an operator requests an MEL, and a Letter of Authorization (LOA) is issued by the FAA, then the use of the MEL becomes mandatory for that helicopter. All maintenance deferrals must be accomplished in accordance with the terms and conditions of the MEL and the operator-generated procedures document.

The use of an MEL for rotorcraft operated under part 91 also allows for the deferral of inoperative items or equipment. The primary guidance becomes the FAA-approved MEL issued to that specific operator and N-numbered helicopter.

The FAA has developed master minimum equipment lists (MMELs) for rotorcraft in current use. Upon written request by a rotorcraft operator, the local FAA Flight Standards District Office (FSDO) may issue the appropriate make and model MMEL, along with an LOA, and the preamble. The operator then develops operations and maintenance (O&M) procedures from the MMEL. This MMEL with O&M procedures now becomes the operator's MEL. The MEL, LOA, preamble, and procedures document developed by the operator must be on board the helicopter when it is operated.

The FAA considers an approved MEL to be a supplemental type certificate (STC) issued to an aircraft by serial number and registration number. It therefore becomes the authority to operate that aircraft in a condition other than originally type certificated.

With an approved MEL, if the position lights were discovered inoperative prior to a daytime flight, the pilot would make an entry in the maintenance record or discrepancy record provided for that purpose. The item is then either repaired or deferred in accordance with the MEL. Upon confirming that daytime flight with inoperative position lights is acceptable in accordance with the provisions of the MEL, the pilot would leave the position lights switch OFF, open the circuit breaker (or whatever action is called for in the procedures document), and placard the position light switch as INOPERATIVE.

There are exceptions to the use of the MEL for deferral. For example, should a component fail that is not listed in the MEL as deferrable (the rotor tachometer, engine tachometer, or cyclic trim, for example), then repairs are required to be performed prior to departure. If maintenance or parts are not readily available at that location, a special flight permit can be obtained from the nearest FSDO. This permit allows the helicopter to be flown to another location for maintenance. This allows an aircraft that may not currently meet applicable airworthiness requirements, but is capable of safe flight, to be operated under the restrictive special terms and conditions attached to the special flight permit.

Deferral of maintenance is not to be taken lightly, and due consideration should be given to the effect an inoperative component may have on the operation of a helicopter, particularly if other items are inoperative. Further information regarding MELs and operations with inoperative equipment can be found in AC 91-67, Minimum Equipment Requirements for General Aviation Operations Under FAR Part 91.

ENGINE START AND ROTOR ENGAGEMENT

During the engine start, rotor engagement, and systems ground check, use the manufacturer's checklists. If a problem arises, have it checked before continuing. Prior to performing these tasks, however, make sure the area near the helicopter is clear of personnel and equipment. Helicopters are safe and efficient flying machines as long as they are operated within the parameters established by the manufacturer.

ROTOR SAFETY CONSIDERATIONS

The exposed nature of the main and tail rotors deserve special caution. You must exercise extreme care when taxiing near hangars or obstructions since the distance between the rotor blade tips and obstructions is very difficult to judge. [Figure 9-1] In addition, you cannot see the tail rotor of some helicopters from the cabin. Therefore, when hovering backwards or turning in those helicopters, allow plenty of room for tail rotor clearance. It is a good practice to glance over your shoulder to maintain this clearance.

Figure 9-1. Exercise extreme caution when hovering near buildings or other aircraft.

Another rotor safety consideration is the thrust a helicopter generates. The main rotor system is capable of blowing sand, dust, snow, ice, and water at high velocities for a significant distance causing injury to nearby people and damage to buildings, automobiles, and other aircraft. Loose snow, can severely reduce visibility and obscure outside visual references. Any airborne debris near the helicopter can be ingested into the engine air intake or struck by the main and tail rotor blades.

SAFETY IN AND AROUND HELICOPTERS

People have been injured, some fatally, in helicopter accidents that would not have occurred had they been informed of the proper method of boarding or deplaning. A properly briefed passenger should never be endangered by a spinning rotor. The simplest method of avoiding accidents of this sort is to stop the rotors before passengers are boarded or allowed to depart. Because this action is not always practicable, and to realize the vast and unique capabilities of the helicopter, it is often necessary to take on passengers or to deplane them while the engine and rotors are turning. To avoid accidents, it is essential that all persons associated with helicopter operations, including passengers, be made aware of all possible hazards and instructed as to how they can be avoided.

Persons directly involved with boarding or deplaning passengers, aircraft servicing, rigging, or hooking up external loads, etc., should be instructed as to their duties. It would be difficult, if not impossible, to cover each and every type of operation related to helicopters. A few of the more obvious and common ones are covered below.

RAMP ATTENDANTS AND AIRCRAFT SERVICING PERSONNEL—These personnel should be instructed as to their specific duties, and the proper method of fulfilling them. In addition, the ramp attendant should be taught to:

1. keep passengers and unauthorized persons out of the helicopter landing and takeoff area.

2. brief passengers on the best way to approach and board a helicopter with its rotors turning.

AIRCRAFT SERVICING—The helicopter rotor blades should be stopped, and both the aircraft and the refueling unit properly grounded prior to any refueling operation. You, as the pilot, should ensure that the proper grade of fuel and the proper additives, when required, are being dispensed.

Refueling the aircraft, while the blades are turning, known as "hot refueling," may be practical for certain types of operation. However, this can be hazardous if not properly conducted. Pilots should remain at the flight controls; and refueling personnel should be knowledgeable about the proper refueling procedures and properly briefed for specific helicopter makes and models.

Refueling units should be positioned to ensure adequate rotor blade clearance. Persons not involved with the refueling operation should keep clear of the area.

Smoking must be prohibited in and around the aircraft during all refueling operations.

EXTERNAL-LOAD RIGGERS—Rigger training is possibly one of the most difficult and continually changing problems of the helicopter external-load operator. A poorly rigged cargo net, light standard, or load pallet could result in a serious and costly accident. It is imperative that all riggers be thoroughly trained to meet the needs of each individual external-load operation. Since rigging requirements may vary several times in a single day, proper training is of the utmost importance to safe operations.

PILOT AT THE FLIGHT CONTROLS—Many helicopter operators have been lured into a "quick turnaround" ground operation to avoid delays at airport terminals and to minimize stop/start cycles of the engine. As part of this quick turnaround, the pilot might leave the cockpit with the engine and rotors turning. Such an operation can be extremely hazardous if a gust of wind disturbs the rotor disc, or the collective flight control moves causing lift to be generated by the rotor system. Either occurrence may cause the helicopter to roll or pitch, resulting in a rotor blade striking the tailboom or the ground. Good operating procedures dictate that pilots remain at the flight controls whenever the engine is running and the rotors are turning.

EXTERNAL-LOAD HOOKUP PERSONNEL—There are several areas in which these personnel should be knowledgeable. First, they should know the lifting capability of the helicopters involved. Since some operators have helicopter models with almost identical physical characteristics but different lifting capabilities, this knowledge is essential. For example, a hookup person may be working with a turbocharged helicopter on a high altitude project when a non-turbocharged helicopter, which looks exactly the same to the ground crew, comes to pick up a load. If the hookup person attaches a load greater than the non-turbocharged helicopter can handle, a potentially dangerous situation could exist.

Second, know the pilots. The safest plan is to standardize all pilots in the manner in which sling loads are picked up and released. Without pilot standardization, the operation could be hazardous. The operator should standardize the pilots on operations while personnel are beneath the helicopter.

Third, know the cargo. Many items carried via sling are very fragile, others can take a beating. The hookup person should always know when a hazardous article is involved and the nature of the hazard, such as explosives, radioactive materials, and toxic chemicals. In addition to knowing this, the hookup person should be familiar with the types of protective gear or clothing and the actions necessary to protect their own safety and that of the operation.

Fourth, know appropriate hand signals. When direct radio communications between ground and flight personnel are not used, the specific meaning of hand signals should be coordinated prior to operations.

Fifth, know emergency procedures. Ground and flight personnel should fully agree to and understand the actions to be taken by all participants in the event of emergencies. This prior planning is essential to avoid injuries to all concerned.

PASSENGERS—All persons who board a helicopter while its rotors are turning should be instructed in the safest means of doing so. Naturally, if you are at the controls, you may not be able to conduct a boarding briefing. Therefore, the individual who arranged for the passengers' flight or is assigned as the ramp attendant should accomplish this task. The exact procedures may vary slightly from one helicopter model to another, but in general the following should suffice.

When boarding—

1. stay away from the rear of the helicopter.

2. approach or leave the helicopter in a crouching manner.

3. approach from the side or front, but never out of the pilot's line of vision.

4. carry tools horizontally, below waist level, never upright or over the shoulder.

5. hold firmly to hats and loose articles.

6. never reach up or dart after a hat or other object that might be blown off or away.

7. protect eyes by shielding them with a hand or by squinting.

8. if suddenly blinded by dust or a blowing object, stop and crouch lower; or better yet, sit down and wait for help.

9. never grope or feel your way toward or away from the helicopter.

Since few helicopters carry cabin attendants, you, as the pilot, will have to conduct the pre-takeoff and pre-landing briefings. The type of operation dictates what sort of briefing is necessary. All briefings should include the following:

1. The use and operation of seatbelts for takeoff, en route, and landing.

2. For overwater flights, the location and use of flotation gear and other survival equipment that might be on board. You should also include how and when to abandon the helicopter should a ditching be necessary.

3. For flights over rough or isolated terrain, all occupants should be told where maps and survival gear are located.

4. Passengers should be instructed as to what actions and precautions to take in the event of an emergency, such as the body position for best spinal protection against a high vertical impact landing (erect with back firmly against the seat back); and when and how to exit after landing. Ensure that passengers are aware of the location of the fire extinguisher and survival equipment.

5. Smoking should not be permitted within 50 feet of an aircraft on the ground. Smoking could be permitted, at the discretion of the pilot, except under the following conditions:

   • during all ground operations.

   • during, takeoff or landing.

   • when carrying flammable or hazardous materials.

When passengers are approaching or leaving a helicopter that is sitting on a slope with the rotors turning, they should approach and depart downhill. This affords the greatest distance between the rotor blades and the ground. If this involves walking around the helicopter, they should always go around the front, never the rear.

VERTICAL TAKEOFF TO A HOVER

A vertical takeoff, or takeoff to a hover, is a maneuver in which the helicopter is raised vertically from the surface to the normal hovering altitude (2 to 5 feet) with a minimum of lateral or longitudinal movement.

TECHNIQUE

Prior to any takeoff or maneuver, you should ensure that the area is clear of other traffic. Then, head the helicopter into the wind, if possible. Place the cyclic in the neutral position, with the collective in the full down position. Increase the throttle smoothly to obtain and maintain proper r.p.m., then raise the collective. Use smooth, continuous movement, coordinating the throttle to maintain proper r.p.m. As you increase the collective, the helicopter becomes light on the skids, and torque tends to cause the nose to swing or yaw to the right unless sufficient left antitorque pedal is used to maintain the heading. (On helicopters with a clockwise main rotor system, the yaw is to the left and right pedal must be applied.)

As the helicopter becomes light on the skids, make necessary cyclic pitch control adjustments to maintain a level attitude. When airborne, use the antitorque pedals to maintain heading and the collective to ensure continuous vertical assent to the normal hovering altitude. When hovering altitude is reached, use the throttle and collective to control altitude, and the cyclic to maintain a stationary hover. Use the antitorque pedals to maintain heading. When a stabilized hover is achieved, check the engine instruments and note the power required to hover. You should also note the position of the cyclic. Cyclic position varies with wind and the amount and distribution of the load.

Excessive movement of any flight control requires a change in the other flight controls. For example, if while hovering, you drift to one side, you naturally move the cyclic in the opposite direction. When you do this, part of the vertical thrust is diverted, resulting in a loss of altitude. To maintain altitude, you must increase the collective. This increases drag on the blades and tends to slow them down. To counteract the drag and maintain r.p.m., you need to increase the throttle. Increased throttle means increased torque, so you must add more pedal pressure to maintain the heading. This can easily lead to overcontrolling the helicopter. However, as your level of proficiency increases, problems associated with overcontrolling decrease.

COMMON ERRORS

1. Failing to ascend vertically as the helicopter becomes airborne.

2. Pulling through on the collective after becoming airborne, causing the helicopter to gain too much altitude.

3. Overcontrolling the antitorque pedals, which not only changes the handling of the helicopter, but also changes the r.p.m.

4. Reducing throttle rapidly in situations where proper r.p.m. has been exceeded. This usually results in exaggerated heading changes and loss of lift, resulting in loss of altitude.

HOVERING

Hovering is a maneuver in which the helicopter is maintained in a nearly motionless flight over a reference point at a constant altitude and on a constant heading. The maneuver requires a high degree of concentration and coordination.

TECHNIQUE

To maintain a hover over a point, you should look for small changes in the helicopter's attitude and altitude. When you note these changes, make the necessary control inputs before the helicopter starts to move from the point. To detect small variations in altitude or position, your main area of visual attention needs to be some distance from the aircraft, using various points on the helicopter or the tip-path plane as a reference. Looking too close or looking down leads to overcontrolling. Obviously, in order to remain over a certain point, you should know where the point is, but your attention should not be focused there.

As with a takeoff, you control altitude with the collective and maintain a constant r.p.m. with the throttle. Use the cyclic to maintain the helicopter's position and the pedals to control heading. To maintain the helicopter in a stabilized hover, make small, smooth, coordinated corrections. As the desired effect occurs, remove the correction in order to stop the helicopter's movement. For example, if the helicopter begins to move rearward, you need to apply a small amount of forward cyclic pressure. However, neutralize this pressure just before the helicopter comes to a stop, or it will begin to move forward.

After you gain experience, you will develop a certain "feel" for the helicopter. You will feel and see small deviations, so you can make the corrections before the helicopter actually moves. A certain relaxed looseness develops, and controlling the helicopter becomes second nature, rather than a mechanical response.

COMMON ERRORS

1. Tenseness and slow reactions to movements of the helicopter.

2. Failure to allow for lag in cyclic and collective pitch, which leads to overcontrolling.

3. Confusing attitude changes for altitude changes, which result in improper use of the controls.

4. Hovering too high, creating a hazardous flight condition.

5. Hovering too low, resulting in occasional touch-down.

HOVERING TURN

A hovering turn is a maneuver performed at hovering altitude in which the nose of the helicopter is rotated either left or right while maintaining position over a reference point on the surface. The maneuver requires the coordination of all flight controls and demands precise control near the surface. You should maintain a constant altitude, rate of turn, and r.p.m.

TECHNIQUE

Initiate the turn in either direction by applying anti-torque pedal pressure toward the desired direction. It should be noted that during a turn to the left, you need to add more power because left pedal pressure increases the pitch angle of the tail rotor, which, in turn, requires additional power from the engine. A turn to the right requires less power. (On helicopters with a clockwise rotating main rotor, right pedal increases the pitch angle and, therefore, requires more power.)

As the turn begins, use the cyclic as necessary (usually into the wind) to keep the helicopter over the desired spot. To continue the turn, you need to add more and more pedal pressure as the helicopter turns to the cross-wind position. This is because the wind is striking the tail surface and tail rotor area, making it more difficult for the tail to turn into the wind. As pedal pressures increase due to crosswind forces, you must increase the cyclic pressure into the wind to maintain position. Use the collective with the throttle to maintain a constant altitude and r.p.m. [Figure 9-2]

After the 90° portion of the turn, you need to decrease pedal pressure slightly to maintain the same rate of turn. Approaching the 180°, or downwind, portion, you need to anticipate opposite pedal pressure due to the tail moving from an upwind position to a down-wind position. At this point, the rate of turn has a tendency to increase at a rapid rate due to the weathervaning tendency of the tail surfaces. Because of the tailwind condition, you need to hold rearward cyclic pressure to keep the helicopter over the same spot.

Because of the helicopter's tendency to weathervane, maintaining the same rate of turn from the 180° position actually requires some pedal pressure opposite the direction of turn. If you do not apply opposite pedal pressure, the helicopter tends to turn at a faster rate. The amount of pedal pressure and cyclic deflection throughout the turn depends on the wind velocity. As you finish the turn on the upwind heading, apply opposite pedal pressure to stop the turn. Gradually apply forward cyclic pressure to keep the helicopter from drifting.

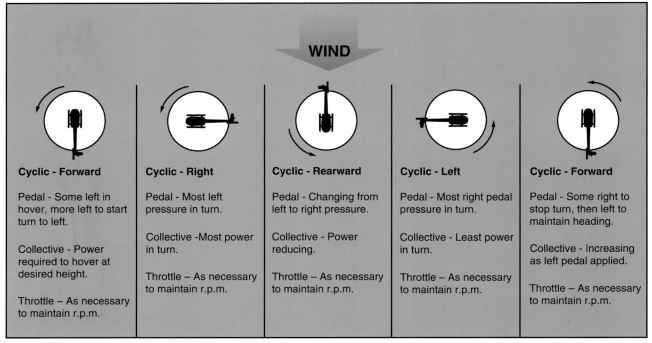

WIND

Cyclic - Forward

Pedal - Some left in hover, more left to start turn to left.

Collective - Power required to hover at desired height.

Throttle – As necessary to maintain r.p.m.

Cyclic - Right

Pedal - Most left pressure in turn.

Collective -Most power in turn.

Throttle – As necessary to maintain r.p.m.

Cyclic - Rearward

Pedal - Changing from left to right pressure.

Collective - Power reducing.

Throttle – As necessary to maintain r.p.m.

Cyclic - Left

Pedal - Most right pedal pressure in turn.

Collective - Least power in turn.

Throttle – As necessary to maintain r.p.m.

Cyclic - Forward

Pedal - Some right to stop turn, then left to maintain heading.

Collective - Increasing as left pedal applied.

Throttle – As necessary to maintain r.p.m.

Figure 9-2. Left turns in helicopters with a counterclockwise rotating main rotor are more difficult to execute because the tail rotor demands more power. This requires that you compensate with additional collective pitch and increased throttle. You might want to refer to this graphic throughout the remainder of the discussion on a hovering turn to the left.

Control pressures and direction of application change continuously throughout the turn. The most dramatic change is the pedal pressure (and corresponding power requirement) necessary to control the rate of turn as the helicopter moves through the downwind portion of the maneuver.

Turns can be made in either direction; however, in a high wind condition, the tail rotor may not be able to produce enough thrust, which means you will not be able to control a turn to the right in a counterclockwise rotor system. Therefore, if control is ever questionable, you should first attempt to make a 90° turn to the left. If sufficient tail rotor thrust exists to turn the helicopter crosswind in a left turn, a right turn can be successfully controlled. The opposite applies to helicopters with clockwise rotor systems. In this case, you should start your turn to the right. Hovering turns should be avoided in winds strong enough to preclude sufficient aft cyclic control to maintain the helicopter on the selected surface reference point when headed downwind. Check the flight manual for the manufacturer's recommendations for this limitation.

COMMON ERRORS

1. Failing to maintain a slow, constant rate of turn.

2. Failing to maintain position over the reference point.

3. Failing to maintain r.p.m. within normal range.

4. Failing to maintain constant altitude.

5. Failing to use the antitorque pedals properly.

HOVERING—FORWARD FLIGHT

You normally use forward hovering flight to move a helicopter to a specific location, and it is usually begun from a stationary hover. During the maneuver, constant groundspeed, altitude, and heading should be maintained.

TECHNIQUE

Before starting, pick out two references directly in front and in line with the helicopter. These reference points should be kept in line throughout the maneuver. [Figure 9-3]

Begin the maneuver from a normal hovering altitude by applying forward pressure on the cyclic. As movement begins, return the cyclic toward the neutral position to keep the groundspeed at a slow rate—no faster than a brisk walk. Throughout the maneuver, maintain a constant groundspeed and path over the ground with the cyclic, a constant heading with the antitorque

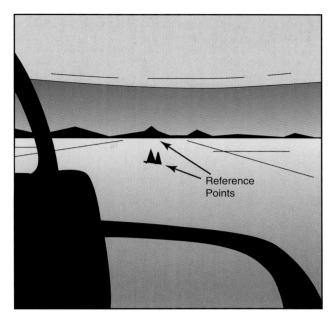

Figure 9-3. To maintain a straight ground track, use two reference points in line and at some distance in front of the helicopter.

pedals, altitude with the collective, and the proper r.p.m. with the throttle.

To stop the forward movement, apply reward cyclic pressure until the helicopter stops. As forward motion stops, return the cyclic to the neutral position to prevent rearward movement. Forward movement can also be stopped by simply applying rearward pressure to level the helicopter and let it drift to a stop.

COMMON ERRORS

1. Exaggerated movement of the cyclic, resulting in erratic movement over the surface.

2. Failure to use the antitorque pedals properly, resulting is excessive heading changes.

3. Failure to maintain desired hovering altitude.

4. Failure to maintain proper r.p.m.

HOVERING—SIDEWARD FLIGHT

Sideward hovering flight may be necessary to move the helicopter to a specific area when conditions make it impossible to use forward flight. During the maneuver, a constant groundspeed, altitude, and heading should be maintained.

TECHNIQUE

Before starting sideward hovering flight, make sure the area you are going to hover into is clear. Then pick two points of reference in a line in the direction of sideward hovering flight to help you maintain the proper ground track. These reference points should be kept in line throughout the maneuver. [Figure 9-4]

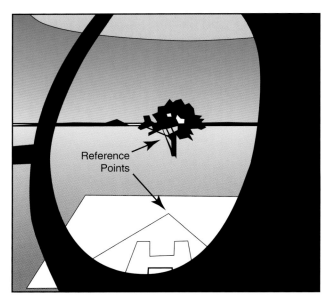

Figure 9-4. The key to hovering sideward is establishing at least two reference points that help you maintain a straight track over the ground while keeping a constant heading.

Begin the maneuver from a normal hovering altitude by applying cyclic toward the side in which the movement is desired. As the movement begins, return the cyclic toward the neutral position to keep the groundspeed at a slow rate—no faster than a brisk walk. Throughout the maneuver, maintain a constant groundspeed and ground track with cyclic. Maintain heading, which in this maneuver is perpendicular to the ground track, with the antitorque pedals, and a constant altitude with the collective. Use the throttle to maintain the proper operating r.p.m.

To stop the sideward movement, apply cyclic pressure in the direction opposite to that of movement and hold it until the helicopter stops. As motion stops, return the cyclic to the neutral position to prevent movement in the opposite direction. Applying sufficient opposite cyclic pressure to level the helicopter may also stop sideward movement. The helicopter then drifts to a stop.

COMMON ERRORS
1. Exaggerated movement of the cyclic, resulting in overcontrolling and erratic movement over the surface.

2. Failure to use proper antitorque pedal control, resulting in excessive heading change.

3. Failure to maintain desired hovering altitude.

4. Failure to maintain proper r.p.m.

5. Failure to make sure the area is clear prior to starting the maneuver.

HOVERING—REARWARD FLIGHT
Rearward hovering flight may be necessary to move the helicopter to a specific area when the situation is such that forward or sideward hovering flight cannot be used. During the maneuver, maintain a constant groundspeed, altitude, and heading. Due to the limited visibility behind a helicopter, it is important that you make sure that the area behind the helicopter is cleared before beginning the maneuver. Use of ground personnel is recommended.

TECHNIQUE
Before starting rearward hovering flight, pick out two reference points in front of, and in line with the helicopter just like you would if you were hovering forward. [Figure 9-3] The movement of the helicopter should be such that these points remain in line.

Begin the maneuver from a normal hovering altitude by applying rearward pressure on the cyclic. After the movement has begun, position the cyclic to maintain a slow groundspeed (no faster than a brisk walk). Throughout the maneuver, maintain constant groundspeed and ground track with the cyclic, a constant heading with the antitorque pedals, constant altitude with the collective, and the proper r.p.m. with the throttle.

To stop the rearward movement, apply forward cyclic and hold it until the helicopter stops. As the motion stops, return the cyclic to the neutral position. Also, as in the case of forward and sideward hovering flight, opposite cyclic can be used to level the helicopter and let it drift to a stop.

COMMON ERRORS
1. Exaggerated movement of the cyclic resulting in overcontrolling and an uneven movement over the surface.

2. Failure to use the antitorque pedals properly, resulting in excessive heading change.

3. Failure to maintain desired hovering altitude.

4. Failure to maintain proper r.p.m.

5. Failure to make sure the area is clear prior to starting the maneuver.

TAXIING
Taxiing refers to operations on, or near the surface of taxiways or other prescribed routes. In helicopters, there are three different types of taxiing.

HOVER TAXI

A "hover taxi" is used when operating below 25 feet AGL. [Figure 9-5] Since hover taxi is just like forward, sideward, or rearward hovering flight, the technique to perform it will not be presented here.

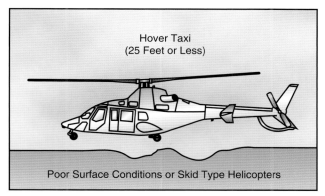

Figure 9-5. Hover taxi.

AIR TAXI

An "air taxi" is preferred when movements require greater distances within an airport or heliport boundary. [Figure 9-6] In this case, you basically fly to your new location; however, you are expected to remain below 100 feet AGL, and to avoid overflight of other aircraft, vehicles, and personnel.

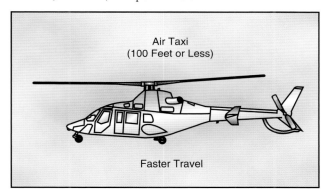

Figure 9-6. Air taxi.

TECHNIQUE

Before starting, determine the appropriate airspeed and altitude combination to remain out of the cross-hatched or shaded areas of the height-velocity diagram. Additionally, be aware of crosswind conditions that could lead to loss of tail rotor effectiveness. Pick out two references directly in front of the helicopter for the ground path desired. These reference points should be kept in line throughout the maneuver.

Begin the maneuver from a normal hovering altitude by applying forward pressure on the cyclic. As movement begins, attain the desired airspeed with the cyclic. Control the desired altitude with the collective, and

r.p.m. with the throttle. Throughout the maneuver, maintain a desired groundspeed and ground track with the cyclic, a constant heading with antitorque pedals, the desired altitude with the collective, and proper operating r.p.m. with the throttle.

To stop the forward movement, apply aft cyclic pressure to reduce forward speed. Simultaneously lower the collective to initiate a descent to hover altitude. As forward motion stops, return the cyclic to the neutral position to prevent rearward movement. When at the proper hover altitude, increase the collective as necessary.

COMMON ERRORS

1. Erratic movement of the cyclic, resulting in improper airspeed control and erratic movement over the surface.

2. Failure to use antitorque pedals properly, resulting in excessive heading changes.

3. Failure to maintain desired altitude.

4. Failure to maintain proper r.p.m.

5. Overflying parked aircraft causing possible damage from rotor downwash.

6. Flying in the cross-hatched or shaded area of the height-velocity diagram.

7. Flying in a crosswind that could lead to loss of tail rotor effectiveness.

SURFACE TAXI

A "surface taxi," for those helicopters with wheels, is used whenever you wish to minimize the effects of rotor downwash. [Figure 9-7]

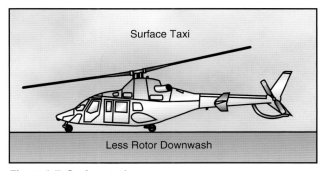

Figure 9-7. Surface taxi.

TECHNIQUE

The helicopter should be in a stationary position on the surface with the collective full down and the r.p.m. the same as that used for a hover. This r.p.m. should be maintained throughout the maneuver. Then, move the cyclic slightly forward and apply gradual upward pressure on the collective to move the helicopter forward along the surface. Use the antitorque pedals to maintain heading and the cyclic to maintain ground track. The collective controls starting, stopping, and speed while taxiing. The higher the collective pitch, the faster the taxi speed; however, you should not taxi faster than a brisk walk. If your helicopter is equipped with brakes, use them to help you slow down. Do not use the cyclic to control groundspeed.

During a crosswind taxi, hold the cyclic into the wind a sufficient amount to eliminate any drifting movement.

COMMON ERRORS
1. Improper use of cyclic.

2. Failure to use antitorque pedals for heading control.

3. Improper use of the controls during crosswind operations.

4. Failure to maintain proper r.p.m.

NORMAL TAKEOFF FROM A HOVER
A normal takeoff from a hover is an orderly transition to forward flight and is executed to increase altitude safely and expeditiously. During the takeoff, fly a profile that avoids the cross-hatched or shaded areas of the height-velocity diagram.

TECHNIQUE
Refer to figure 9-8 (position 1). Bring the helicopter to a hover and make a performance check, which includes power, balance, and flight controls. The power check should include an evaluation of the amount of excess power available; that is, the difference between the power being used to hover and the power available at the existing altitude and temperature conditions. The balance condition of the helicopter is indicated by the position of the cyclic when maintaining a stationary hover. Wind will necessitate some cyclic deflection, but there should not be an extreme deviation from neutral. Flight controls must move freely, and the helicopter should respond normally. Then visually clear the area all around.

Start the helicopter moving by smoothly and slowly easing the cyclic forward (position 2). As the helicopter starts to move forward, increase the collective, as necessary, to prevent the helicopter from sinking and adjust the throttle to maintain r.p.m. The increase in power requires an increase in the proper antitorque pedal to maintain heading. Maintain a straight takeoff path throughout the takeoff. As you accelerate through effective translational lift (position 3), the helicopter begins to climb and the nose tends to rise due to increased lift. At this point adjust the collective to obtain normal climb power and apply enough forward cyclic to overcome the tendency of the nose to rise. At position 4, hold an attitude that allows a smooth acceleration toward climbing airspeed and a commensurate gain in altitude so that the takeoff profile does not take you through any of the cross-hatched or shaded areas of the height-velocity diagram. As airspeed increases (position 5), the streamlining of the fuselage reduces engine torque effect, requiring a gradual reduction of antitorque pedal pressure. As the helicopter continues to climb and accelerate to best rate of climb, apply aft cyclic pressure to raise the nose smoothly to the normal climb attitude.

COMMON ERRORS
1. Failing to use sufficient collective pitch to prevent loss of altitude prior to attaining translational lift.

2. Adding power too rapidly at the beginning of the transition from hovering to forward flight without forward cyclic compensation, causing the helicopter to gain excessive altitude before acquiring airspeed.

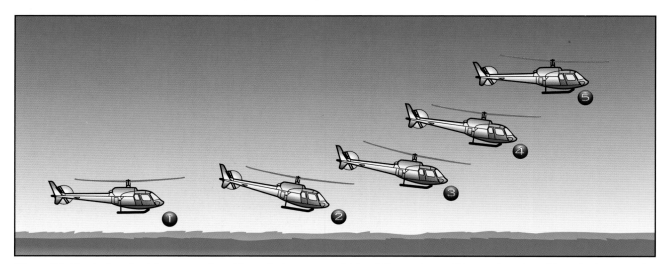

Figure 9-8. The helicopter takes several positions during a normal takeoff from a hover. The numbered positions in the text refer to the numbers in this illustration.

3. Assuming an extreme nose-down attitude near the surface in the transition from hovering to forward flight.

4. Failing to maintain a straight flight path over the surface (ground track).

5. Failing to maintain proper airspeed during the climb.

6. Failing to adjust the throttle to maintain proper r.p.m.

NORMAL TAKEOFF FROM THE SURFACE

Normal takeoff from the surface is used to move the helicopter from a position on the surface into effective translational lift and a normal climb using a minimum amount of power. If the surface is dusty or covered with loose snow, this technique provides the most favorable visibility conditions and reduces the possibility of debris being ingested by the engine.

TECHNIQUE

Place the helicopter in a stationary position on the surface. Lower the collective to the full down position, and reduce the r.p.m. below operating r.p.m. Visually clear the area and select terrain features, or other objects, to aid in maintaining the desired track during takeoff and climb out. Increase the throttle to the proper r.p.m. and raise the collective slowly until the helicopter is light on the skids. Hesitate momentarily and adjust the cyclic and antitorque pedals, as necessary, to prevent any surface movement. Continue to apply upward collective and, as the helicopter breaks ground, use the cyclic, as necessary, to begin forward movement as altitude is gained. Continue to accelerate, and as effective translational lift is attained, the helicopter begins to climb. Adjust attitude and power, if necessary, to climb in the same manner as a takeoff from a hover.

COMMOM ERRORS

1. Departing the surface in an attitude that is too nose-low. This situation requires the use of excessive power to initiate a climb.

2. Using excessive power combined with a level attitude, which causes a vertical climb.

3. Too abrupt application of the collective when departing the surface, causing r.p.m. and heading control errors.

CROSSWIND CONSIDERATIONS DURING TAKEOFFS

If the takeoff is made during crosswind conditions, the helicopter is flown in a slip during the early stages of the maneuver. [Figure 9-9] The cyclic is held into the wind a sufficient amount to maintain the desired ground track for the takeoff. The heading is maintained with the use of the antitorque pedals. In other words, the rotor is tilted into the wind so that the sideward movement of the helicopter is just enough to counteract the crosswind effect. To prevent the nose from turning in the direction of the rotor tilt, it is necessary to increase the antitorque pedal pressure on the side opposite the rotor tilt.

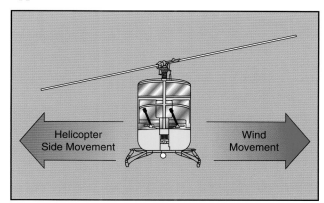

Figure 9-9. During a slip, the rotor disc is tilted into the wind.

After approximately 50 feet of altitude is gained, make a coordinated turn into the wind to maintain the desired ground track. This is called crabbing into the wind. The stronger the crosswind, the more you have to turn the helicopter into the wind to maintain the desired ground track. [Figure 9-10]

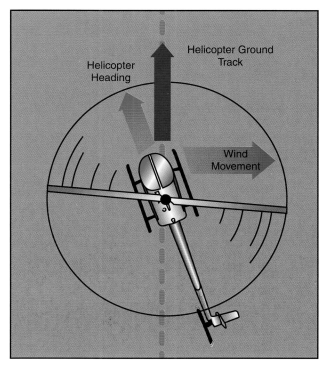

Figure 9-10. To compensate for wind drift at altitude, crab the helicopter into the wind.

STRAIGHT-AND-LEVEL FLIGHT

Straight-and-level flight is flight in which a constant altitude and heading are maintained. The attitude of the helicopter determines the airspeed and is controlled by the cyclic. Altitude is primarily controlled by use of the collective.

TECHNIQUE

To maintain forward flight, the rotor tip-path plane must be tilted forward to obtain the necessary horizontal thrust component from the main rotor. This generally results in a nose-low attitude. The lower the nose, the greater the power required to maintain altitude, and the higher the resulting airspeed. Conversely, the greater the power used, the lower the nose must be to maintain altitude. [Figure 9-11]

Figure 9-11. You can maintain a straight-and-level attitude by keeping the tip-path plane parallel to and a constant distance above or below the natural horizon. For any given airspeed, this distance remains the same as long as you sit in the same position in the same type of aircraft.

When in straight-and-level flight, any increase in the collective, while holding airspeed constant, causes the helicopter to climb. A decrease in the collective, while holding airspeed constant, causes the helicopter to descend. A change in the collective requires a coordinated change of the throttle to maintain a constant r.p.m. Additionally, the antitorque pedals need to be adjusted to maintain heading and to keep the helicopter in longitudinal trim.

To increase airspeed in straight-and-level flight, apply forward pressure on the cyclic and raise the collective as necessary to maintain altitude. To decrease airspeed, apply rearward pressure on the cyclic and lower the collective, as necessary, to maintain altitude.

Although the cyclic is sensitive, there is a slight delay in control reaction, and it will be necessary to anticipate actual movement of the helicopter. When making cyclic inputs to control the altitude or airspeed of a helicopter, take care not to overcontrol. If the nose of the helicopter rises above the level-flight attitude, apply forward pressure to the cyclic to bring the nose down. If this correction is held too long, the nose drops too low. Since the helicopter continues to change attitude momentarily after the controls reach neutral, return the

cyclic to neutral slightly before the desired attitude is reached. This principal holds true for any cyclic input.

Since helicopters are inherently unstable, if a gust or turbulence causes the nose to drop, the nose tends to continue to drop instead of returning to a straight-and-level attitude as would a fixed-wing aircraft. Therefore, you must remain alert and FLY the helicopter at all times.

COMMON ERRORS

1. Failure to properly trim the helicopter, tending to hold antitorque pedal pressure and opposite cyclic. This is commonly called cross-controlling.

2. Failure to maintain desired airspeed.

3. Failure to hold proper control position to maintain desired ground track.

TURNS

A turn is a maneuver used to change the heading of the helicopter. The aerodynamics of a turn were previously discussed in Chapter 3—Aerodynamics of Flight.

TECHNIQUE

Before beginning any turn, the area in the direction of the turn must be cleared not only at the helicopter's altitude, but also above and below. To enter a turn from straight-and-level flight, apply sideward pressure on the cyclic in the direction the turn is to be made. This is the only control movement needed to start the turn. Do not use the pedals to assist the turn. Use the pedals only to compensate for torque to keep the helicopter in longitudinal trim. [Figure 9-12]

How fast the helicopter banks depends on how much lateral cyclic pressure you apply. How far the helicopter banks (the steepness of the bank) depends on how long you displace the cyclic. After establishing the proper bank angle, return the cyclic toward the neutral position. Increase the collective and throttle to main-

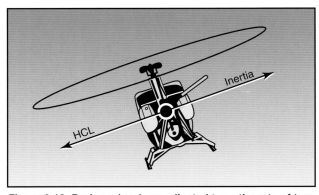

Figure 9-12. During a level, coordinated turn, the rate of turn is commensurate with the angle of bank used, and inertia and horizontal component of lift (HCL) are equal.

tain altitude and r.p.m. As the torque increases, increase the proper antitorque pedal pressure to maintain longitudinal trim. Depending on the degree of bank, additional forward cyclic pressure may be required to maintain airspeed.

Rolling out of the turn to straight-and-level flight is the same as the entry into the turn except that pressure on the cyclic is applied in the opposite direction. Since the helicopter continues to turn as long as there is any bank, start the rollout before reaching the desired heading.

The discussion on level turns is equally applicable to making turns while climbing or descending. The only difference being that the helicopter is in a climbing or descending attitude rather than that of level flight. If a simultaneous entry is desired, merely combine the techniques of both maneuvers—climb or descent entry and turn entry. When recovering from a climbing or descending turn, the desired heading and altitude are rarely reached at the same time. If the heading is reached first, stop the turn and maintain the climb or descent until reaching the desired altitude. On the other hand, if the altitude is reached first, establish the level flight attitude and continue the turn to the desired heading.

SLIPS
A slip occurs when the helicopter slides sideways toward the center of the turn. [Figure 9-13] It is caused by an insufficient amount of antitorque pedal in the direction of the turn, or too much in the direction opposite the turn, in relation to the amount of power used. In other words, if you hold improper antitorque pedal pressure, which keeps the nose from following the turn, the helicopter slips sideways toward the center of the turn.

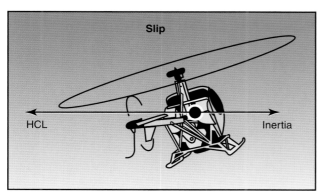
Figure 9-13. During a slip, the rate of turn is too slow for the angle of bank used, and the horizontal component of lift (HCL) exceeds inertia.

SKIDS
A skid occurs when the helicopter slides sideways away from the center of the turn. [Figure 9-14] It is caused by too much antitorque pedal pressure in the direction of the turn, or by too little in the direction

opposite the turn in relation to the amount of power used. If the helicopter is forced to turn faster with increased pedal pressure instead of by increasing the degree of the bank, it skids sideways away from the center of the turn instead of flying in its normal curved pattern.

In summary, a skid occurs when the rate of turn is too fast for the amount of bank being used, and a slip occurs when the rate of turn is too slow for the amount of bank being used.

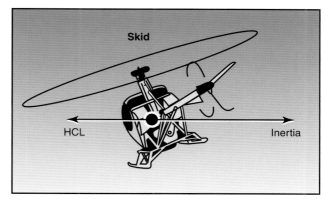
Figure 9-14. During a skid, the rate of turn is too fast for the angle of bank used, and inertia exceeds the horizontal component of lift (HCL).

COMMON ERRORS
1. Using antitorque pedal pressures for turns. This is usually not necessary for small helicopters.

2. Slipping or skidding in the turn.

NORMAL CLIMB
The entry into a climb from a hover has already been discussed under "Normal Takeoff from a Hover;" therefore, this discussion is limited to a climb entry from cruising flight.

TECHNIQUE
To enter a climb from cruising flight, apply aft cyclic to obtain the approximate climb attitude. Simultaneously increase the collective and throttle to obtain climb power and maintain r.p.m. In a counterclockwise rotor system, increase the left antitorque pedal pressure to compensate for the increased torque. As the airspeed approaches normal climb airspeed, adjust the cyclic to hold this airspeed. Throughout the maneuver, maintain climb attitude, heading, and airspeed with the cyclic; climb power and r.p.m. with the collective and throttle; and longitudinal trim with the antitorque pedals.

To level off from a climb, start adjusting the attitude to the level flight attitude a few feet prior to reaching the desired altitude. The amount of lead depends on the rate of climb at the time of level-off (the higher the rate of climb, the more the lead). Generally, the lead is 10 percent of the climb rate. For example, if your climb rate is 500 feet per minute, you should lead the level-off by 50 feet.

To begin the level-off, apply forward cyclic to adjust and maintain a level flight attitude, which is slightly nose low. You should maintain climb power until the airspeed approaches the desired cruising airspeed, then lower the collective to obtain cruising power and adjust the throttle to obtain and maintain cruising r.p.m. Throughout the level-off, maintain longitudinal trim and heading with the antitorque pedals.

COMMON ERRORS
1. Failure to maintain proper power and airspeed.

2. Holding too much or too little antitorque pedal.

3. In the level-off, decreasing power before lowering the nose to cruising attitude.

NORMAL DESCENT
A normal descent is a maneuver in which the helicopter loses altitude at a controlled rate in a controlled attitude.

TECHNIQUE
To establish a normal descent from straight-and-level flight at cruising airspeed, lower the collective to obtain proper power, adjust the throttle to maintain r.p.m., and increase right antitorque pedal pressure to maintain heading in a counterclockwise rotor system, or left pedal pressure in a clockwise system. If cruising airspeed is the same as, or slightly above descending airspeed, simultaneously apply the necessary cyclic pressure to obtain the approximate descending attitude. If cruising speed is well above descending airspeed, you can maintain a level flight attitude until the airspeed approaches the descending airspeed, then lower the nose to the descending attitude. Throughout the maneuver, maintain descending attitude and airspeed with the cyclic; descending power and r.p.m. with the collective and throttle; and heading with the antitorque pedals.

To level off from the descent, lead the desired altitude by approximately 10 percent of the rate of descent. For example, a 500 feet per minute rate of descent would require a 50 foot lead. At this point, increase the collective to obtain cruising power, adjust the throttle to maintain r.p.m., and increase left antitorque pedal pressure to maintain heading (right pedal pressure in a clockwise rotor system). Adjust the cyclic to obtain cruising airspeed and a level flight attitude as the desired altitude is reached.

COMMON ERRORS
1. Failure to maintain constant angle of decent during training.

2. Failure to lead the level-off sufficiently, which results in recovery below the desired altitude.

3. Failure to adjust antitorque pedal pressures for changes in power.

GROUND REFERENCE MANEUVERS
Ground reference maneuvers are training exercises flown to help you develop a division of attention between the flight path and ground references, while controlling the helicopter and watching for other aircraft in the vicinity. Prior to each maneuver, a clearing turn should be accomplished to ensure the practice area is free of conflicting traffic.

RECTANGULAR COURSE
The rectangular course is a training maneuver in which the ground track of the helicopter is equidistant from all sides of a selected rectangular area on the ground. While performing the maneuver, the altitude and airspeed should be held constant. The rectangular course helps you to develop a recognition of a drift toward or away from a line parallel to the intended ground track. This is helpful in recognizing drift toward or from an airport runway during the various legs of the airport traffic pattern.

For this maneuver, pick a square or rectangular field, or an area bounded on four sides by section lines or roads, where the sides are approximately a mile in length. The area selected should be well away from other air traffic. Fly the maneuver approximately 600 to 1,000 feet above the ground, which is the altitude usually required for an airport traffic pattern. You should fly the helicopter parallel to and at a uniform distance, about one-fourth to one-half mile, from the field boundaries, not above the boundaries. For best results, position your flight path outside the field boundaries just far enough away that they may be easily observed from either pilot seat by looking out the side of the helicopter. If an attempt is made to fly directly above the edges of the field, you will have no usable reference points to start and complete the turns. In addition, the closer the track of the helicopter is to the field boundaries, the steeper the bank necessary at the turning points. Also, you should be able to see the edges of the selected field while seated in a normal position and looking out the side of the helicopter during either a left-hand or right-hand course. The distance of the ground track from the edges of the field should be the same regardless of whether the course is flown to the left or right. All turns should be started when your helicopter is abeam the corners of the field boundaries. The bank normally should not exceed 30°.

Although the rectangular course may be entered from any direction, this discussion assumes entry on a

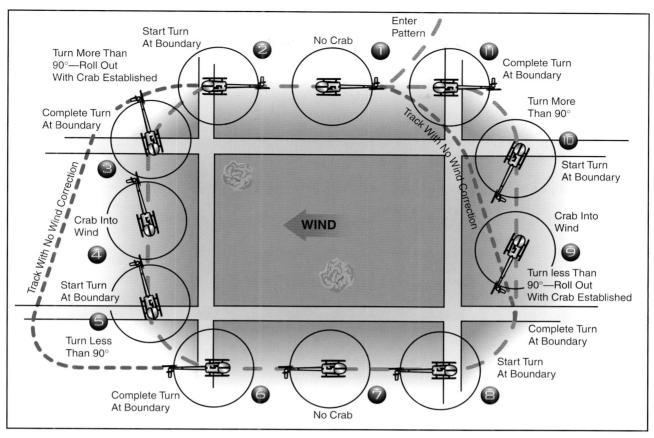

Figure 9-15. Rectangular course. The numbered positions in the text refer to the numbers in this illustration.

downwind heading. [Figure 9-15] As you approach the field boundary on the downwind leg, you should begin planning for your turn to the crosswind leg. Since you have a tailwind on the downwind leg, the helicopter's groundspeed is increased (position 1). During the turn onto the crosswind leg, which is the equivalent of the base leg in a traffic pattern, the wind causes the helicopter to drift away from the field. To counteract this effect, the roll-in should be made at a fairly fast rate with a relatively steep bank (position 2).

As the turn progresses, the tailwind component decreases, which decreases the groundspeed. Consequently, the bank angle and rate of turn must be reduced gradually to ensure that upon completion of the turn, the crosswind ground track continues to be the same distance from the edge of the field. Upon completion of the turn, the helicopter should be level and aligned with the downwind corner of the field. However, since the crosswind is now pushing you away from the field, you must establish the proper drift correction by flying slightly into the wind. Therefore, the turn to crosswind should be greater than a 90° change in heading (position 3). If the turn has been made properly, the field boundary again appears to be one-fourth to one-half mile away. While on the crosswind leg, the wind correction should be adjusted, as

necessary, to maintain a uniform distance from the field boundary (position 4).

As the next field boundary is being approached (position 5), plan the turn onto the upwind leg. Since a wind correction angle is being held into the wind and toward the field while on the crosswind leg, this next turn requires a turn of less than 90°. Since the crosswind becomes a headwind, causing the groundspeed to decrease during this turn, the bank initially must be medium and progressively decreased as the turn proceeds. To complete the turn, time the rollout so that the helicopter becomes level at a point aligned with the corner of the field just as the longitudinal axis of the helicopter again becomes parallel to the field boundary (position 6). The distance from the field boundary should be the same as on the other sides of the field.

On the upwind leg, the wind is a headwind, which results in an decreased groundspeed (position 7). Consequently, enter the turn onto the next leg with a fairly slow rate of roll-in, and a relatively shallow bank (position 8). As the turn progresses, gradually increase the bank angle because the headwind component is diminishing, resulting in an increasing groundspeed. During and after the turn onto this leg, the wind tends to drift the helicopter toward the field boundary. To compensate for the drift, the amount of turn must be less than 90° (position 9).

Again, the rollout from this turn must be such that as the helicopter becomes level, the nose of the helicopter is turned slightly away the field and into the wind to correct for drift. The helicopter should again be the same distance from the field boundary and at the same altitude, as on other legs. Continue the crosswind leg until the downwind leg boundary is approached (position 10). Once more you should anticipate drift and turning radius. Since drift correction was held on the crosswind leg, it is necessary to turn greater than 90° to align the helicopter parallel to the downwind leg boundary. Start this turn with a medium bank angle, gradually increasing it to a steeper bank as the turn progresses. Time the rollout to assure paralleling the boundary of the field as the helicopter becomes level (position 11).

If you have a direct headwind or tailwind on the upwind and downwind leg, drift should not be encountered. However, it may be difficult to find a situation where the wind is blowing exactly parallel to the field boundaries. This makes it necessary to use a slight wind correction angle on all the legs. It is important to anticipate the turns to compensate for groundspeed, drift, and turning radius. When the wind is behind the helicopter, the turn is faster and steeper; when it is ahead of the helicopter, the turn is slower and shallower. These same techniques apply while flying in an airport traffic pattern.

S-TURNS

Another training maneuver you might use is the S-turn, which helps you correct for wind drift in turns. This maneuver requires turns to the left and right. The reference line used, whether a road, railroad, or fence, should be straight for a considerable distance and should extend as nearly perpendicular to the wind as possible.

The object of S-turns is to fly a pattern of two half circles of equal size on opposite sides of the reference line. [Figure 9-16] The maneuver should be performed at a constant altitude between 600 and 1,000 feet above the terrain. S-turns may be started at any point; however, during early training it may be beneficial to start on a downwind heading. Entering downwind permits the immediate selection of the steepest bank that is desired throughout the maneuver. The discussion that follows is based on choosing a reference line that is perpendicular to the wind and starting the maneuver on a downwind heading.

As the helicopter crosses the reference line, immediately establish a bank. This initial bank is the steepest used throughout the maneuver since the helicopter is headed directly downwind and the groundspeed is at its

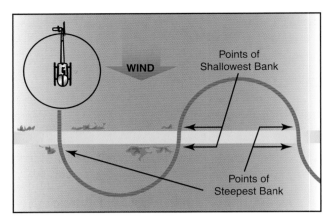

Figure 9-16. S-turns across a road.

highest. Gradually reduce the bank, as necessary, to describe a ground track of a half circle. Time the turn so that as the rollout is completed, the helicopter is crossing the reference line perpendicular to it and heading directly upwind. Immediately enter a bank in the opposite direction to begin the second half of the "S." Since the helicopter is now on an upwind heading, this bank (and the one just completed before crossing the reference line) is the shallowest in the maneuver. Gradually increase the bank, as necessary, to describe a ground track that is a half circle identical in size to the one previously completed on the other side of the reference line. The steepest bank in this turn should be attained just prior to rollout when the helicopter is approaching the reference line nearest the downwind heading. Time the turn so that as the rollout is complete, the helicopter is perpendicular to the reference line and is again heading directly downwind.

In summary, the angle of bank required at any given point in the maneuver is dependent on the groundspeed. The faster the groundspeed, the steeper the bank; the slower the groundspeed, the shallower the bank. To express it another way, the more nearly the helicopter is to a downwind heading, the steeper the bank; the more nearly it is to an upwind heading, the shallower the bank. In addition to varying the angle of bank to correct for drift in order to maintain the proper radius of turn, the helicopter must also be flown with a drift correction angle (crab) in relation to its ground track; except of course, when it is on direct upwind or downwind headings or there is no wind. One would normally think of the fore and aft axis of the helicopter as being tangent to the ground track pattern at each point. However, this is not the case. During the turn on the upwind side of the reference line (side from which the wind is blowing), crab the nose of the helicopter toward the outside of the circle. During the turn on the downwind side of the reference line (side of the reference line opposite to the direction from which the wind is blowing), crab the nose of the helicopter toward the inside of the circle. In either case, it is obvious that the helicopter is being crabbed into the wind just as it is when trying to maintain a straight ground track. The amount of crab depends upon the wind velocity and how nearly the helicopter is to a crosswind position. The stronger the wind, the greater the crab angle at any given position for a turn of a given radius. The more nearly the helicopter is to a crosswind position, the greater the crab angle. The maximum crab angle should be at the point of each half circle farthest from the reference line.

A standard radius for S-turns cannot be specified, since the radius depends on the airspeed of the helicopter, the velocity of the wind, and the initial bank chosen for entry.

TURNS AROUND A POINT

This training maneuver requires you to fly constant radius turns around a preselected point on the ground using a bank of approximately 30°, while maintaining a constant altitude. [Figure 9-17] Your objective, as in other ground reference maneuvers, is to develop the ability to subconsciously control the helicopter while dividing attention between the flight path and ground references, while still watching for other air traffic in the vicinity.

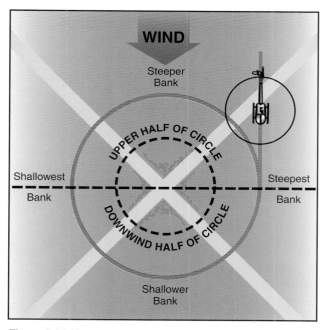

Figure 9-17. Turns around a point.

The factors and principles of drift correction that are involved in S-turns are also applicable in this maneuver. As in other ground track maneuvers, a constant radius around a point will, if any wind exists, require a constantly changing angle of bank and angles of wind correction. The closer the helicopter is to a direct downwind heading where the groundspeed is

greatest, the steeper the bank, and the faster the rate of turn required to establish the proper wind correction angle. The more nearly it is to a direct upwind heading where the groundspeed is least, the shallower the bank, and the slower the rate of turn required to establish the proper wind correction angle. It follows, then, that throughout the maneuver, the bank and rate of turn must be gradually varied in proportion to the groundspeed.

The point selected for turns around a point should be prominent and easily distinguishable, yet small enough to present a precise reference. Isolated trees, crossroads, or other similar small landmarks are usually suitable. The point should be in an area away from communities, livestock, or groups of people on the ground to prevent possible annoyance or hazard to others. Since the maneuver is performed between 600 and 1,000 feet AGL, the area selected should also afford an opportunity for a safe emergency autorotation in the event it becomes necessary.

To enter turns around a point, fly the helicopter on a downwind heading to one side of the selected point at a distance equal to the desired radius of turn. When any significant wind exists, it is necessary to roll into the initial bank at a rapid rate so that the steepest bank is attained abeam the point when the helicopter is headed directly downwind. By entering the maneuver while heading directly downwind, the steepest bank can be attained immediately. Thus, if a bank of 30° is desired, the initial bank is 30° if the helicopter is at the correct distance from the point. Thereafter, the bank is gradually shallowed until the point is reached where the helicopter is headed directly upwind. At this point, the bank is gradually steepened until the steepest bank is again attained when heading downwind at the initial point of entry.

Just as S-turns require that the helicopter be turned into the wind in addition to varying the bank, so do turns around a point. During the downwind half of the circle, the helicopter's nose must be progressively turned toward the inside of the circle; during the upwind half, the nose must be progressively turned toward the outside. The downwind half of the turn around the point may be compared to the downwind side of the S-turn, while the upwind half of the turn around a point may be compared to the upwind side of the S-turn.

As you become experienced in performing turns around a point and have a good understanding of the effects of wind drift and varying of the bank angle and wind correction angle as required, entry into the maneuver may be from any point. When entering this maneuver at any point, the radius of the turn must be carefully selected, taking into account the wind velocity and groundspeed so that an excessive bank is not required later on to maintain the proper ground track.

COMMON ERRORS DURING GROUND REFERENCE MANEUVERS

1. Faulty entry technique.

2. Poor planning, orientation, or division of attention.

3. Uncoordinated flight control application.

4. Improper correction for wind drift.

5. An unsymmetrical ground track during S-Turns Across a Road.

6. Failure to maintain selected altitude or airspeed.

7. Selection of a ground reference where there is no suitable emergency landing area within gliding distance.

TRAFFIC PATTERNS

A traffic pattern is useful to control the flow of traffic, particularly at airports without operating control towers. It affords a measure of safety, separation, protection, and administrative control over arriving, departing, and circling aircraft. Due to specialized operating characteristics, airplanes and helicopters do not mix well in the same traffic environment. At multiple-use airports, you routinely must avoid the flow of fixed-wing traffic. To do this, you need to be familiar with the patterns typically flown by airplanes. In addition, you should learn how to fly these patterns in case air traffic control (ATC) requests that you fly a fixed-wing traffic pattern.

A normal traffic pattern is rectangular, has five named legs, and a designated altitude, usually 600 to 1,000 feet AGL. A pattern in which all turns are to the left is called a standard pattern. [Figure 9-18] The takeoff leg (item 1) normally consists of the aircraft's flight path after takeoff. This leg is also called the upwind leg. You should turn to the crosswind leg (item 2), after passing the departure end of the runway when you are at a safe altitude. Fly the downwind leg (item 3) parallel to the runway at the designated traffic pattern altitude and distance from the runway. Begin the base leg (item 4) at a point selected according to other traffic and wind conditions. If the wind is very strong, begin the turn sooner than normal. If the wind is light, delay the turn to base. The final approach (item 5) is the path the aircraft flies immediately prior to touchdown.

You may find variations at different localities and at airports with operating control towers. For example, a right-hand pattern may be designated to expedite the

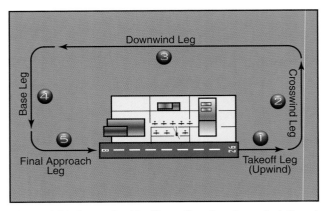

Figure 9-18. A standard traffic pattern has turns to left and five designated legs.

flow of traffic when obstacles or highly populated areas make the use of a left-hand pattern undesirable.

When approaching an airport with an operating control tower in a helicopter, it is possible to expedite traffic by stating your intentions, for example:

1. (Call sign of helicopter) Robinson 8340J.

2. (Position) 10 miles west.

3. (Request) for landing and hover to...

In order to avoid the flow of fixed-wing traffic, the tower will often clear you direct to an approach point or to a particular runway intersection nearest your destination point. At uncontrolled airports, if at all possible, you should adhere to standard practices and patterns.

Traffic pattern entry procedures at airports with an operating control tower are specified by the controller. At uncontrolled airports, traffic pattern altitudes and entry procedures may vary according to established local procedures. The general procedure is for you to enter the pattern at a 45° angle to the downwind leg abeam the midpoint of the runway. For information concerning traffic pattern and landing direction, you should utilize airport advisory service or UNICOM, when available.

The standard departure procedure when using the fixed-wing traffic pattern is usually straight-out, downwind, or a right-hand departure. When a control tower is in operation, you can request the type of departure you desire. In most cases, helicopter departures are made into the wind unless obstacles or traffic dictate otherwise. At airports without an operating control tower, you must comply with the departure procedures established for that airport.

APPROACHES

An approach is the transition from traffic pattern altitude to either a hover or to the surface. The approach should terminate at the hover altitude with the rate of descent and groundspeed reaching zero at the same time. Approaches are categorized according to the angle of descent as normal, steep, or shallow. In this chapter we will concentrate on the normal approach. Steep and shallow approaches are discussed in the next chapter.

You should use the type of approach best suited to the existing conditions. These conditions may include obstacles, size and surface of the landing area, density altitude, wind direction and speed, and weight. Regardless of the type of approach, it should always be made to a specific, predetermined landing spot.

NORMAL APPROACH TO A HOVER

A normal approach uses a descent profile of between 8° and 12° starting at approximately 300 feet AGL.

TECHNIQUE

On final approach, at the recommended approach airspeed and at approximately 300 feet AGL, align the helicopter with the point of intended touchdown. [Figure 9-19] After intercepting an approach angle of 8° to 12°, begin the approach by lowering the collective sufficiently to get the helicopter decelerating and descending down the approach angle. With the decrease in the collective, the nose tends to pitch down, requiring aft cyclic to maintain the recommended approach airspeed attitude. Adjust antitorque pedals, as necessary, to maintain longitudinal trim. You can determine the proper approach angle by relating the point of intended touchdown to a point on the helicopter windshield. The collective controls the angle of approach. If the touchdown point seems to be moving up on the windshield, the angle is becoming shallower, necessitating a slight increase in collective. If the touchdown point moves down on the windshield, the approach angle is becoming steeper, requiring a slight decrease in collective. Use the cyclic to control the rate of closure or how fast your are moving toward the touchdown point. Maintain entry airspeed until the apparent groundspeed and rate of closure appear to be increasing. At this point, slowly begin decelerating with slight aft cyclic, and smoothly lower the collective to maintain approach angle. Use the cyclic to maintain a rate of closure equivalent to a brisk walk.

At approximately 25 to 40 feet AGL, depending on wind, the helicopter begins to lose effective translational lift. To compensate for loss of effective translational lift, you must increase the collective to maintain the approach angle, while maintaining the proper r.p.m. The increase of collective pitch tends to make the nose rise, requiring forward cyclic to maintain the proper rate of closure.

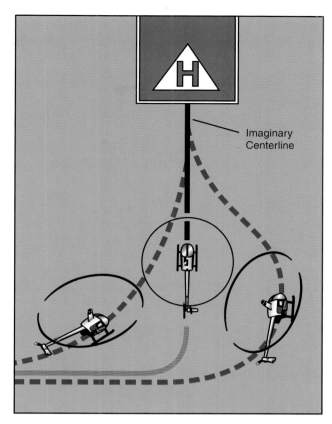

Figure 9-19. Plan the turn to final so the helicopter rolls out on an imaginary extension of the centerline for the final approach path. This path should neither angle to the landing area, as shown by the helicopter on the left, nor require an S-turn, as shown by the helicopter on the right.

As the helicopter approaches the recommended hover altitude, you need to increase the collective sufficiently to maintain the hover. At the same time you need to apply aft cyclic to stop any forward movement, while controlling the heading with antitorque pedals.

COMMON ERRORS

1. Failing to maintain proper r.p.m. during the entire approach.

2. Improper use of the collective in controlling the angle of descent.

3. Failing to make antitorque pedal corrections to compensate for collective changes during the approach.

4. Failing to simultaneously arrive at hovering altitude and attitude with zero groundspeed.

5. Low r.p.m. in transition to the hover at the end of the approach.

6. Using too much aft cyclic close to the surface, which may result in tail rotor strikes.

NORMAL APPROACH TO THE SURFACE

A normal approach to the surface or a no-hover landing is used if loose snow or dusty surface conditions exist. These situations could cause severely restricted visibility, or the engine could possibly ingest debris when the helicopter comes to a hover. The approach is the same as the normal approach to a hover; however, instead of terminating at a hover, continue the approach to touchdown. Touchdown should occur with the skids level, zero groundspeed, and a rate of descent approaching zero.

TECHNIQUE:

As the helicopter nears the surface, increase the collective, as necessary, to cushion the landing on the surface, terminate in a skids-level attitude with no forward movement.

COMMON ERRORS

1. Terminating at a hover, then making a vertical landing.

2. Touching down with forward movement.

3. Approaching too slow, requiring the use of excessive power during the termination.

4. Approaching too fast, causing a hard landing.

CROSSWIND DURING APPROACHES

During a crosswind approach, you should crab into the wind. At approximately 50 feet of altitude, use a slip to align the fuselage with the ground track. The rotor is tilted into the wind with cyclic pressure so that the sideward movement of the helicopter and wind drift counteract each other. Maintain the heading and ground track with the antitorque pedals. This technique should be used on any type of crosswind approach, whether it is a shallow, normal, or steep approach.

GO-AROUND

A go-around is a procedure for remaining airborne after an intended landing is discontinued. A go-around may be necessary when:

- Instructed by the control tower.

- Traffic conflict occurs.

A good rule of thumb to use during an approach is to make a go-around if the helicopter is in a position from which it is not safe to continue the approach. Anytime you feel an approach is uncomfortable, incorrect, or potentially dangerous, abandon the approach. The decision to make a go-around should be positive and initiated before a critical situation develops. When the decision is made, carry it out without hesitation. In most cases, when you initiate the go-around, power is at a low setting. Therefore, your first response is to increase collective to takeoff power. This movement is coordinated with the throttle to maintain r.p.m., and the proper antitorque pedal to control heading. Then, establish a climb attitude and maintain climb speed to go around for another approach.

AFTER LANDING AND SECURING

When the flight is terminated, park the helicopter where it will not interfere with other aircraft and not be a hazard to people during shutdown. Rotor downwash can cause damage to other aircraft in close proximity, and spectators may not realize the danger or see the rotors turning. Passengers should remain in the helicopter with their seats belts secured until the rotors have stopped turning. During the shutdown and postflight inspection, follow the manufacturer's checklist. Any discrepancies found should be noted and, if necessary, reported to maintenance personnel.

NOISE ABATEMENT PROCEDURES

The FAA, in conjunction with airport operators and community leaders, is now using noise abatement procedures to reduce the level of noise generated by aircraft departing over neighborhoods that are near airports. The airport authority may simply request that you use a designated runway, wind permitting. You also may be asked to restrict some of your operations, such as practicing landings, during certain time periods. There are three ways to determine the noise abatement procedure at an airport. First, if there is a control tower on the field, they will assign the preferred noise abatement runway or takeoff direction to you. Second, you can check the *Airport/Facility Directory* for information on local procedures. Third, there may be information for you to read in the pilot's lounge, or even signs posted next to a runway that will advise you on local procedures.

CHAPTER 10

Advanced Flight Maneuvers

The maneuvers presented in this chapter require more finesse and understanding of the helicopter and the surrounding environment. When performing these maneuvers, you will probably be taking your helicopter to the edge of the safe operating envelope. Therefore, if you are ever in doubt about the outcome of the maneuver, you should abort the mission entirely or wait for more favorable conditions.

RECONNAISSANCE PROCEDURES

Anytime you are planning to land or takeoff at an unfamiliar site, you should gather as much information as you can about the area. Reconnaissance techniques are ways of gathering this information.

HIGH RECONNAISSANCE

The purpose of a high reconnaissance is to determine the wind direction and speed, a point for touchdown, the suitability of the landing area, the approach and departure axes, obstacles and their effect on wind patterns, and the most suitable flight paths into and out of the area. When conducting a high reconnaissance, give particular consideration to forced landing areas in case of an emergency.

Altitude, airspeed, and flight pattern for a high reconnaissance are governed by wind and terrain features. You must strike a balance between a reconnaissance conducted too high and one too low. It should not be flown so low that you have to divide your attention between studying the area and avoiding obstructions to flight. A high reconnaissance should be flown at an altitude of 300 to 500 feet above the surface. A general rule to follow is to ensure that sufficient altitude is available at all times to land into the wind in case of engine failure. In addition, a 45° angle of observation generally allows the best estimate of the height of barriers, the presence of obstacles, the size of the area, and the slope of the terrain. Always maintain safe altitudes and airspeeds, and keep a forced landing area within reach whenever possible.

LOW RECONNAISSANCE

A low reconnaissance is accomplished during the approach to the landing area. When flying the approach, verify what was observed in the high reconnaissance, and check for anything new that may have been missed at a higher altitude, such as wires, slopes, and small crevices. If everything is alright, you can complete the approach to a landing. However, you must make the decision to land or go-around before effective translational lift is lost.

If a decision is made to complete the approach, terminate it in a hover, so you can carefully check the landing point before lowering the helicopter to the surface. Under certain conditions, it may be desirable to continue the approach to the surface. Once the helicopter is on the ground, maintain operating r.p.m. until you have checked the stability of the helicopter to be sure it is in a secure and safe position.

GROUND RECONNAISSANCE

Prior to departing an unfamiliar location, make a detailed analysis of the area. There are several factors to consider during this evaluation. Besides determining the best departure path, you must select a route that will get your helicopter from its present position to the takeoff point.

Some things to consider while formulating a takeoff plan are the aircraft load, height of obstacles, the shape of the area, and direction of the wind. If the helicopter is heavily loaded, you must determine if there is sufficient power to clear the obstacles. Sometimes it is better to pick a path over shorter obstacles than to take off directly into the wind. You should also evaluate the shape of the area so that you can pick a path that will give you the most room to maneuver and abort the takeoff if necessary. Wind analysis also helps determine the route of takeoff. The prevailing wind can be altered by obstructions on the departure path, and can significantly affect aircraft performance. One way to determine the wind direction is to drop some dust or grass, and observe which way it is blowing. Keep in mind that if the main rotor is turning, you will need to be a sufficient distance from the helicopter to ensure that the downwash of the blades does not give you a false indication.

If possible, you should walk the route from the helicopter to the takeoff position. Evaluate obstacles that could be hazardous and ensure that you will have adequate rotor clearance. Once at the downwind end of the available area, mark a position for takeoff so that the tail and main rotors have sufficient clearance from any obstructions behind the helicopter. Use a sturdy marker, such as a heavy stone or log, so it does not blow away.

MAXIMUM PERFORMANCE TAKEOFF

A maximum performance takeoff is used to climb at a steep angle to clear barriers in the flight path. It can be used when taking off from small areas surrounded by high obstacles. Before attempting a maximum performance takeoff, you must know thoroughly the capabilities and limitations of your equipment. You must also consider the wind velocity, temperature, altitude, gross weight, center-of-gravity location, and other factors affecting your technique and the performance of the helicopter.

To safely accomplish this type of takeoff, there must be enough power to hover, in order to prevent the helicopter from sinking back to the surface after becoming airborne. This hover power check can be used to determine if there is sufficient power available to accomplish this maneuver.

The angle of climb for a maximum performance takeoff depends on existing conditions. The more critical the conditions, such as high density altitudes, calm winds, and high gross weights, the shallower the angle of climb. In light or no wind conditions, it might be necessary to operate in the crosshatched or shaded areas of the height/velocity diagram during the beginning of this maneuver. Therefore, be aware of the calculated risk when operating in these areas. An engine failure at a low altitude and airspeed could place the helicopter in a dangerous position, requiring a high degree of skill in making a safe autorotative landing.

TECHNIQUE

Before attempting a maximum performance takeoff, bring the helicopter to a hover, and determine the excess power available by noting the difference between the power available and that required to hover. You should also perform a balance and flight control check and note the position of the cyclic. Then position the helicopter into the wind and return the helicopter to the surface. Normally, this maneuver is initiated from the surface. After checking the area for obstacles and other aircraft, select reference points along the takeoff path to maintain ground track. You should also consider alternate routes in case you are not able to complete the maneuver. [Figure 10-1]

Begin the takeoff by getting the helicopter light on the skids (position 1). Pause and neutralize all aircraft movement. Slowly increase the collective and position the cyclic so as to break ground in a 40 knot attitude. This is approximately the same attitude as when the helicopter is light on the skids. Continue to slowly increase the collective until the maximum power available is reached. This large collective movement requires a substantial increase in pedal pressure to maintain heading (position 2). Use the cyclic, as necessary, to control movement toward the desired flight path and, therefore, climb angle during the maneuver (position 3). Maintain rotor r.p.m. at its maximum, and do not allow it to decrease since you would probably have to lower the collective to regain it. Maintain these inputs until the helicopter clears the obstacle, or until reaching 50 feet for demonstration purposes (position 4). Then, establish a normal climb attitude and reduce power (position 5). As in any maximum performance maneuver, the techniques you use affect the actual results. Smooth, coordinated inputs coupled with precise control allow the helicopter to attain its maximum performance.

COMMON ERRORS

1. Failure to consider performance data, including height/velocity diagram.

2. Nose too low initially, causing horizontal flight rather than more vertical flight.

3. Failure to maintain maximum permissible r.p.m.

4. Abrupt control movements.

5. Failure to resume normal climb power and airspeed after clearing the obstacle.

RUNNING/ROLLING TAKEOFF

A running takeoff in a skid-type helicopter or a rolling takeoff in a wheeled helicopter is sometimes used when conditions of load and/or density altitude prevent a sustained hover at normal hovering altitude. However, you should not attempt this maneuver if you do not have sufficient power to hover, at least momentarily. If the helicopter cannot be hovered, its performance is unpredictable. If the helicopter cannot be raised off the surface at all, sufficient power might not be available to safely accomplish the maneuver. If you cannot momentarily hover the helicopter, you must wait for conditions to improve or off-load some of the weight.

To accomplish a safe running or rolling takeoff, the surface area must be of sufficient length and smoothness, and there cannot be any barriers in the flight path to interfere with a shallow climb.

For wheeled helicopters, a rolling takeoff is sometimes used to minimize the downwash created during a takeoff from a hover.

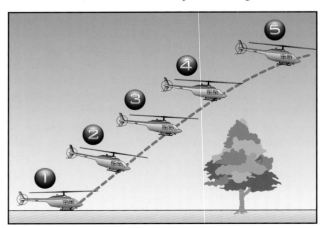

Figure 10-1. Maximum performance takeoff.

TECHNIQUE

Refer to figure 10-2. To begin the maneuver, first align the helicopter to the takeoff path. Next, increase the throttle to obtain takeoff r.p.m., and increase the collective smoothly until the helicopter becomes light on the skids or landing gear (position 1). Then, move the cyclic slightly forward of the neutral hovering position, and apply additional collective to start the forward movement (position 2). To simulate a reduced power condition during practice, use one to two inches less manifold pressure, or three to five percent less torque, than that required to hover.

Figure 10-2. Running/rolling takeoff.

Maintain a straight ground track with lateral cyclic and heading with antitorque pedals until a climb is established. As effective translational lift is gained, the helicopter becomes airborne in a fairly level attitude with little or no pitching (position 3). Maintain an altitude to take advantage of ground effect, and allow the airspeed to increase toward normal climb speed. Then, follow a climb profile that takes you through the clear area of the height/velocity diagram (position 4). During practice maneuvers, after you have climbed to an altitude of 50 feet, establish the normal climb power setting and attitude.

COMMON ERRORS

1. Failing to align heading and ground track to keep surface friction to a minimum.

2. Attempting to become airborne before obtaining effective translational lift.

3. Using too much forward cyclic during the surface run.

4. Lowering the nose too much after becoming airborne, resulting in the helicopter settling back to the surface.

5. Failing to remain below the recommended altitude until airspeed approaches normal climb speed.

RAPID DECELERATION (QUICK STOP)

In normal operations, use the rapid deceleration or quick stop maneuver to slow the helicopter rapidly and bring it to a stationary hover. The maneuver requires a high degree of coordination of all controls. It is practiced at an altitude that permits a safe clearance between the tail rotor and the surface throughout the maneuver, especially at the point where the pitch attitude is highest. The altitude at completion should be no higher than the maximum safe hovering altitude prescribed by the manufacturer. In selecting an altitude at which to begin the maneuver, you should take into account the overall length of the helicopter and the height/velocity diagram. Even though the maneuver is called a rapid deceleration or quick stop, it is performed slowly and smoothly with the primary emphasis on coordination.

TECHNIQUE

During training always perform this maneuver into the wind. [Figure 10-3, position 1] After leveling off at an altitude between 25 and 40 feet, depending on the manufacturer's recommendations, accelerate to the desired entry speed, which is approximately 45 knots for most training helicopters (position 2). The altitude you choose should be high enough to avoid danger to the tail rotor during the flare, but low enough to stay out of the crosshatched or shaded areas of the height/velocity diagram throughout the maneuver. In addition, this altitude should be low enough that you can bring the helicopter to a hover during the recovery.

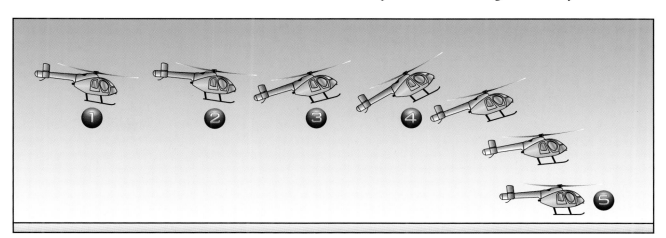

Figure 10-3. Rapid deceleration or quick stop.

At position 3, initiate the deceleration by applying aft cyclic to reduce forward speed. Simultaneously, lower the collective, as necessary, to counteract any climbing tendency. The timing must be exact. If you apply too little down collective for the amount of aft cyclic applied, a climb results. If you apply too much down collective, a descent results. A rapid application of aft cyclic requires an equally rapid application of down collective. As collective pitch is lowered, apply proper antitorque pedal pressure to maintain heading, and adjust the throttle to maintain r.p.m.

After attaining the desired speed (position 4), initiate the recovery by lowering the nose and allowing the helicopter to descend to a normal hovering altitude in level flight and zero groundspeed (position 5). During the recovery, increase collective pitch, as necessary, to stop the helicopter at normal hovering altitude, adjust the throttle to maintain r.p.m., and apply proper pedal pressure, as necessary, to maintain heading.

COMMON ERRORS

1. Initiating the maneuver by applying down collective.

2. Initially applying aft cyclic stick too rapidly, causing the helicopter to **balloon**.

3. Failing to effectively control the rate of deceleration to accomplish the desired results.

4. Allowing the helicopter to stop forward motion in a tail-low attitude.

5. Failing to maintain proper r.p.m.

6. Waiting too long to apply collective pitch (power) during the recovery, resulting in excessive manifold pressure or an over-torque situation when collective pitch is applied rapidly.

7. Failing to maintain a safe clearance over the terrain.

8. Improper use of antitorque pedals resulting in erratic heading changes.

STEEP APPROACH TO A HOVER
A steep approach is used primarily when there are obstacles in the approach path that are too high to allow a normal approach. A steep approach permits entry into most confined areas and is sometimes used to avoid areas of turbulence around a pinnacle. An approach angle of approximately 15° is considered a steep approach. [Figure 10-4]

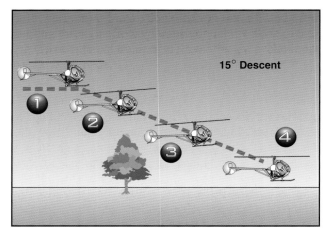

Figure 10-4. Steep approach to a hover.

TECHNIQUE
On final approach, head your helicopter into the wind and align it with the intended touchdown point at the recommended approach airspeed (position 1). When you intercept an approach angle of 15°, begin the approach by lowering the collective sufficiently to start the helicopter descending down the approach path and decelerating (position 2). Use the proper antitorque pedal for trim. Since this angle is steeper than a normal approach angle, you need to reduce the collective more than that required for a normal approach. Continue to decelerate with slight aft cyclic, and smoothly lower the collective to maintain the approach angle. As in a normal approach, reference the touchdown point on the windshield to determine changes in approach angle. This point is in a lower position than a normal approach. Aft cyclic is required to decelerate sooner than a normal approach, and the rate of closure becomes apparent at a higher altitude. Maintain the approach angle and rate of descent with the collective, rate of closure with the cyclic, and trim with antitorque pedals. Use a crab above 50 feet and a slip below 50 feet for any crosswind that might be present.

Loss of effective translational lift occurs higher in a steep approach (position 3), requiring an increase in the collective to prevent settling, and more forward cyclic to achieve the proper rate of closure. Terminate the approach at hovering altitude above the intended landing point with zero groundspeed (position 4). If power has been properly applied during the final portion of the approach, very little additional power is required in the hover.

Balloon—Gaining an excessive amount of altitude as a result of an abrupt flare.

COMMON ERRORS

1. Failing to maintain proper r.p.m. during the entire approach.

2. Improper use of collective in maintaining the selected angle of descent.

3. Failing to make antitorque pedal corrections to compensate for collective pitch changes during the approach.

4. Slowing airspeed excessively in order to remain on the proper angle of descent.

5. Inability to determine when effective translational lift is lost.

6. Failing to arrive at hovering altitude and attitude, and zero groundspeed almost simultaneously.

7. Low r.p.m. in transition to the hover at the end of the approach.

8. Using too much aft cyclic close to the surface, which may result in the tail rotor striking the surface.

SHALLOW APPROACH AND RUNNING/ROLL-ON LANDING

Use a shallow approach and running landing when a high-density altitude or a high gross weight condition, or some combination thereof, is such that a normal or steep approach cannot be made because of insufficient power to hover. [Figure 10-5] To compensate for this lack of power, a shallow approach and running landing makes use of translational lift until surface contact is made. If flying a wheeled helicopter, you can also use a roll-on landing to minimize the effect of downwash. The glide angle for a shallow approach is approximately 5°. Since the helicopter will be sliding or rolling to a stop during this maneuver, the landing area must be smooth and long enough to accomplish this task.

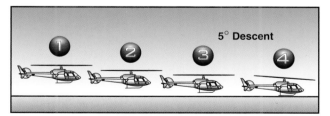

Figure 10-5. Shallow approach and running landing.

TECHNIQUE

A shallow approach is initiated in the same manner as the normal approach except that a shallower angle of descent is maintained. The power reduction to initiate the desired angle of descent is less than that for a normal approach since the angle of descent is less (position 1).

As you lower the collective, maintain heading with proper antitorque pedal pressure, and r.p.m. with the throttle. Maintain approach airspeed until the apparent rate of closure appears to be increasing. Then, begin to slow the helicopter with aft cyclic (position 2).

As in normal and steep approaches, the primary control for the angle and rate of descent is the collective, while the cyclic primarily controls the groundspeed. However, there must be a coordination of all the controls for the maneuver to be accomplished successfully. The helicopter should arrive at the point of touchdown at or slightly above effective translational lift. Since translational lift diminishes rapidly at slow airspeeds, the deceleration must be smoothly coordinated, at the same time keeping enough lift to prevent the helicopter from settling abruptly.

Just prior to touchdown, place the helicopter in a level attitude with the cyclic, and maintain heading with the antitorque pedals. Use the cyclic to keep the heading and ground track identical (position 3). Allow the helicopter to descend gently to the surface in a straight-and-level attitude, cushioning the landing with the collective. After surface contact, move the cyclic slightly forward to ensure clearance between the tailboom and the rotor disc. You should also use the cyclic to maintain the surface track. (position 4). You normally hold the collective stationary until the helicopter stops; however, if you want more braking action, you can lower the collective slightly. Keep in mind that due to the increased ground friction when you lower the collective, the helicopter's nose might pitch forward. Exercise caution not to correct this pitching movement with aft cyclic since this movement could result in the rotor making contact with the tailboom. During the landing, maintain normal r.p.m. with the throttle and directional control with the antitorque pedals.

For wheeled helicopters, use the same technique except after landing, lower the collective, neutralize the controls, and apply the brakes, as necessary, to slow the helicopter. Do not use aft cyclic when bringing the helicopter to a stop.

COMMON ERRORS

1. Assuming excessive nose-high attitude to slow the helicopter near the surface.

2. Insufficient collective and throttle to cushion landing.

3. Failing to add proper antitorque pedal as collective is added to cushion landing, resulting in a touchdown while the helicopter is moving sideward.

4. Failing to maintain a speed that takes advantage of effective translational lift.

5. Touching down at an excessive groundspeed for the existing conditions. (Some helicopters have maximum touchdown groundspeeds.)

6. Failing to touch down in a level attitude.

7. Failing to maintain proper r.p.m. during and after touchdown.

8. Poor directional control during touchdown.

SLOPE OPERATIONS

Prior to conducting any slope operations, you should be thoroughly familiar with the characteristics of dynamic rollover and mast bumping, which are discussed in Chapter 11—Helicopter Emergencies. The approach to a slope is similar to the approach to any other landing area. During slope operations, make allowances for wind, barriers, and forced landing sites in case of engine failure. Since the slope may constitute an obstruction to wind passage, you should anticipate turbulence and downdrafts.

SLOPE LANDING

You usually land a helicopter across the slope rather than with the slope. Landing with the helicopter facing down the slope or downhill is not recommended because of the possibility of striking the tail rotor on the surface.

TECHNIQUE

Refer to figure 10-6. At the termination of the approach, move the helicopter slowly toward the slope, being careful not to turn the tail upslope. Position the helicopter across the slope at a stabilized hover headed into the wind over the spot of intended landing (frame 1). Downward pressure on the collective starts the helicopter descending. As the upslope skid touches the ground, hesitate momentarily in a level attitude, then apply lateral cyclic in the direction of the slope (frame 2). This holds the skid against the slope while you continue lowering the downslope skid with the collective. As you lower the collective, continue to move the cyclic toward the slope to maintain a fixed position (frame 3). The slope must be shallow enough so you

can hold the helicopter against it with the cyclic during the entire landing. A slope of 5° is considered maximum for normal operation of most helicopters.

You should be aware of any abnormal vibration or mast bumping that signals maximum cyclic deflection. If this occurs, abandon the landing because the slope is too steep. In most helicopters with a counterclockwise rotor system, landings can be made on steeper slopes when you are holding the cyclic to the right. When landing on slopes using left cyclic, some cyclic input must be used to overcome the translating tendency. If wind is not a factor, you should consider the drifting tendency when determining landing direction.

After the downslope skid is on the surface, reduce the collective to full down, and neutralize the cyclic and pedals (frame 4). Normal operating r.p.m. should be maintained until the full weight of the helicopter is on the landing gear. This ensures adequate r.p.m. for immediate takeoff in case the helicopter starts sliding down the slope. Use antitorque pedals as necessary throughout the landing for heading control. Before reducing the r.p.m., move the cyclic control as necessary to check that the helicopter is firmly on the ground.

COMMON ERRORS

1. Failure to consider wind effects during the approach and landing.

2. Failure to maintain proper r.p.m. throughout the entire maneuver.

3. Turning the tail of the helicopter into the slope.

4. Lowering the downslope skid or wheel too rapidly.

5. Applying excessive cyclic control into the slope, causing mast bumping.

SLOPE TAKEOFF

A slope takeoff is basically the reverse of a slope landing. [Figure 10-7] Conditions that may be associated with the slope, such as turbulence and obstacles, must

Figure 10-6. Slope landing.

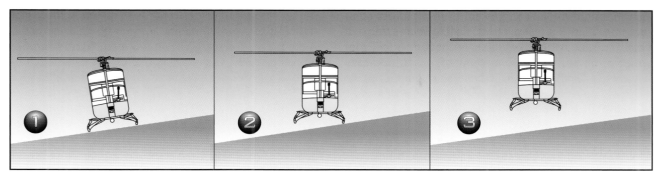

Figure 10-7. Slope takeoff.

be considered during the takeoff. Planning should include suitable forced landing areas.

TECHNIQUE

Begin the takeoff by increasing r.p.m. to the normal range with the collective full down. Then, move the cyclic toward the slope (frame 1). Holding cyclic toward the slope causes the downslope skid to rise as you slowly raise the collective (frame 2). As the skid comes up, move the cyclic toward the neutral position. If properly coordinated, the helicopter should attain a level attitude as the cyclic reaches the neutral position. At the same time, use antitorque pedal pressure to maintain heading and throttle to maintain r.p.m. With the helicopter level and the cyclic centered, pause momentarily to verify everything is correct, and then gradually raise the collective to complete the liftoff (frame 3).

After reaching a hover, take care to avoid hitting the ground with the tail rotor. If an upslope wind exists, execute a crosswind takeoff and then make a turn into the wind after clearing the ground with the tail rotor.

COMMON ERRORS

1. Failure to adjust cyclic control to keep the helicopter from sliding downslope.

2. Failure to maintain proper r.p.m.

3. Holding excessive cyclic into the slope as the downslope skid is raised.

4. Turning the tail of the helicopter into the slope during takeoff.

CONFINED AREA OPERATIONS

A confined area is an area where the flight of the helicopter is limited in some direction by terrain or the presence of obstructions, natural or manmade. For example, a clearing in the woods, a city street, a road, a building roof, etc., can each be regarded as a confined area. Generally, takeoffs and landings should be made into the wind to obtain maximum airspeed with minimum groundspeed.

There are several things to consider when operating in confined areas. One of the most important is maintaining a clearance between the rotors and obstacles forming the confined area. The tail rotor deserves special consideration because, in some helicopters, you cannot always see it from the cabin. This not only applies while making the approach, but while hovering as well. Another consideration is that wires are especially difficult to see; however, their supporting devices, such as poles or towers, serve as an indication of their presence and approximate height. If any wind is present, you should also expect some turbulence. [Figure 10-8]

Figure 10-8. If the wind velocity is 10 knots or greater, you should expect updrafts on the windward side and downdrafts on the lee side of obstacles. You should plan the approach with these factors in mind, but be ready to alter your plans if the wind speed or direction changes.

Something else for you to consider is the availability of forced landing areas during the planned approach. You should think about the possibility of flying from one alternate landing area to another throughout the approach, while avoiding unfavorable areas. Always leave yourself a way out in case the landing cannot be completed or a go-around is necessary.

APPROACH

A high reconnaissance should be completed before initiating the confined area approach. Start the approach phase using the wind and speed to the best possible advantage. Keep in mind areas suitable for forced landing. It may be necessary to choose between an approach that is crosswind, but over an open area, and one directly into the wind, but over heavily wooded or extremely rough terrain where a safe forced landing would be impossible. If these conditions exist, consider the possibility of making the initial phase of the approach crosswind over the open area and then turning into the wind for the final portion of the approach.

Always operate the helicopter as close to its normal capabilities as possible, taking into consideration the situation at hand. In all confined area operations, with the exception of the pinnacle operation, the angle of descent should be no steeper than necessary to clear any barrier in the approach path and still land on the selected spot. The angle of climb on takeoff should be normal, or not steeper than necessary to clear any barrier. Clearing a barrier by a few feet and maintaining normal operating r.p.m., with perhaps a reserve of power, is better than clearing a barrier by a wide margin but with a dangerously low r.p.m. and no power reserve.

Always make the landing to a specific point and not to some general area. This point should be located well forward, away from the approach end of the area. The more confined the area, the more essential it is that you land the helicopter precisely at a definite point. Keep this point in sight during the entire final approach.

When flying a helicopter near obstructions, always consider the tail rotor. A safe angle of descent over barriers must be established to ensure tail rotor clearance of all obstructions. After coming to a hover, take care to avoid turning the tail into obstructions.

TAKEOFF

A confined area takeoff is considered an **altitude over airspeed** maneuver. Before takeoff, make a ground reconnaissance to determine the type of takeoff to be performed, to determine the point from which the takeoff should be initiated to ensure the maximum amount of available area, and finally, how to best maneuver the helicopter from the landing point to the proposed takeoff position.

If wind conditions and available area permit, the helicopter should be brought to a hover, turned around, and hovered forward from the landing position to the takeoff position. Under certain conditions, sideward flight to the takeoff position may be necessary. If rearward

flight is required to reach the takeoff position, place reference markers in front of the helicopter in such a way that a ground track can be safely followed to the takeoff position. In addition, the takeoff marker should be located so that it can be seen without hovering beyond it.

When planning the takeoff, consider the direction of the wind, obstructions, and forced landing areas. To help you fly up and over an obstacle, you should form an imaginary line from a point on the leading edge of the helicopter to the highest obstacle to be cleared. Fly this line of ascent with enough power to clear the obstacle by a safe distance. After clearing the obstacle, maintain the power setting and accelerate to the normal climb speed. Then, reduce power to the normal climb power setting.

COMMON ERRORS

1. Failure to perform, or improper performance of, a high or low reconnaissance.

2. Flying the approach angle at too steep or too shallow an approach for the existing conditions.

3. Failing to maintain proper r.p.m.

4. Failure to consider emergency landing areas.

5. Failure to select a specific landing spot.

6. Failure to consider how wind and turbulence could affect the approach.

7. Improper takeoff and climb technique for existing conditions.

PINNACLE AND RIDGELINE OPERATIONS

A pinnacle is an area from which the surface drops away steeply on all sides. A ridgeline is a long area from which the surface drops away steeply on one or two sides, such as a bluff or precipice. The absence of obstacles does not necessarily lessen the difficulty of pinnacle or ridgeline operations. Updrafts, downdrafts, and turbulence, together with unsuitable terrain in which to make a forced landing, may still present extreme hazards.

APPROACH AND LANDING

If you need to climb to a pinnacle or ridgeline, do it on the upwind side, when practicable, to take advantage of any updrafts. The approach flight path should be parallel to the ridgeline and into the wind as much as possible. [Figure 10-9]

Load, altitude, wind conditions, and terrain features determine the angle to use in the final part of an approach. As a general rule, the greater the winds, the steeper the approach needs to be to avoid turbulent air and downdrafts. Groundspeed during the approach is

---

Altitude over Airspeed—In this type of maneuver, it is more important to gain altitude than airspeed. However, unless operational considerations dictate otherwise, the crosshatched or shaded areas of the height/velocity diagram should be avoided.

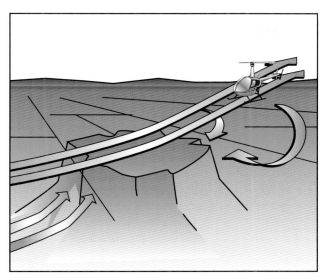

Figure 10-9. When flying an approach to a pinnacle or ridgeline, avoid the areas where downdrafts are present, especially when excess power is limited. If you encounter downdrafts, it may become necessary to make an immediate turn away from the pinnacle to avoid being forced into the rising terrain.

more difficult to judge because visual references are farther away than during approaches over trees or flat terrain. If a crosswind exists, remain clear of downdrafts on the leeward or downwind side of the ridgeline. If the wind velocity makes the crosswind landing hazardous, you may be able to make a low, coordinated turn into the wind just prior to terminating the approach. When making an approach to a pinnacle, avoid leeward turbulence and keep the helicopter within reach of a forced landing area as long as possible.

On landing, take advantage of the long axis of the area when wind conditions permit. Touchdown should be made in the forward portion of the area. Always perform a stability check, prior to reducing r.p.m., to ensure the landing gear is on firm terrain that can safely support the weight of the helicopter.

TAKEOFF

A pinnacle takeoff is an **airspeed over altitude** maneuver made from the ground or from a hover. Since pinnacles and ridgelines are generally higher than the immediate surrounding terrain, gaining airspeed on the takeoff is more important than gaining altitude. The higher the airspeed, the more rapid the departure from slopes of the pinnacle. In addition to covering unfavorable terrain rapidly, a higher airspeed affords a more favorable glide angle and thus contributes to the chances of reaching a safe area in the event of a forced landing. If a suitable forced landing area is not available, a higher airspeed also permits a more effective flare prior to making an autorotative landing.

On takeoff, as the helicopter moves out of ground effect, maintain altitude and accelerate to normal climb airspeed. When normal climb speed is attained, establish a normal climb attitude. Never dive the helicopter down the slope after clearing the pinnacle.

COMMON ERRORS

1. Failure to perform, or improper performance of, a high or low reconnaissance.

2. Flying the approach angle at too steep or too shallow an approach for the existing conditions.

3. Failure to maintain proper r.p.m.

4. Failure to consider emergency landing areas.

5. Failure to consider how wind and turbulence could affect the approach and takeoff.

Airspeed over Altitude—This means that in this maneuver, obstacles are not a factor, and it is more important to gain airspeed than altitude.

Helicopter Emergencies

Today helicopters are quite reliable. However emergencies do occur, whether a result of mechanical failure or pilot error. By having a thorough knowledge of the helicopter and its systems, you will be able to more readily handle the situation. In addition, by knowing the conditions that can lead to an emergency, many potential accidents can be avoided.

AUTOROTATION

In a helicopter, an autorotation is a descending maneuver where the engine is disengaged from the main rotor system and the rotor blades are driven solely by the upward flow of air through the rotor. In other words, the engine is no longer supplying power to the main rotor.

The most common reason for an autorotation is an engine failure, but autorotations can also be performed in the event of a complete tail rotor failure, since there is virtually no torque produced in an autorotation. If altitude permits, they can also be used to recover from settling with power. If the engine fails, the freewheeling unit automatically disengages the engine from the main rotor allowing the main rotor to rotate freely. Essentially, the freewheeling unit disengages anytime the engine r.p.m. is less than the rotor r.p.m.

At the instant of engine failure, the main rotor blades are producing lift and thrust from their angle of attack and velocity. By immediately lowering collective pitch, which must be done in case of an engine failure, lift and drag are reduced, and the helicopter begins an immediate descent, thus producing an upward flow of air through the rotor system. This upward flow of air through the rotor provides sufficient thrust to maintain rotor r.p.m. throughout the descent. Since the tail rotor is driven by the main rotor transmission during autorotation, heading control is maintained as in normal flight.

Several factors affect the rate of descent in autorotation; density altitude, gross weight, rotor r.p.m., and airspeed. Your primary control of the rate of descent is airspeed. Higher or lower airspeeds are obtained with the cyclic pitch control just as in normal flight. In theory, you have a choice in the angle of descent varying from a vertical descent to maximum range, which is the minimum angle of descent. Rate of descent is high at zero airspeed and decreases to a minimum at approximately 50 to 60 knots, depending upon the particular helicopter and the factors just mentioned. As the airspeed increases beyond that which gives minimum rate of descent, the rate of descent increases again.

When landing from an autorotation, the energy stored in the rotating blades is used to decrease the rate of descent and make a soft landing. A greater amount of rotor energy is required to stop a helicopter with a high rate of descent than is required to stop a helicopter that is descending more slowly. Therefore, autorotative descents at very low or very high airspeeds are more critical than those performed at the minimum rate of descent airspeed.

Each type of helicopter has a specific airspeed at which a power-off glide is most efficient. The best airspeed is the one which combines the greatest glide range with the slowest rate of descent. The specific airspeed is somewhat different for each type of helicopter, yet certain factors affect all configurations in the same manner. For specific autorotation airspeeds for a particular helicopter, refer to the FAA-approved rotorcraft flight manual.

The specific airspeed for autorotations is established for each type of helicopter on the basis of average weather and wind conditions and normal loading. When the helicopter is operated with heavy loads in high density altitude or gusty wind conditions, best performance is achieved from a slightly increased airspeed in the descent. For autorotations at low density altitude and light loading, best performance is achieved from a slight decrease in normal airspeed. Following this general procedure of fitting airspeed to existing conditions, you can achieve approximately the same glide angle in any set of circumstances and estimate the touchdown point.

When making turns during an autorotation, generally use cyclic control only. Use of antitorque pedals to assist or speed the turn causes loss of airspeed and downward pitching of the nose. When an autorotation is initiated, sufficient antitorque pedal pressure should be used to maintain straight flight and prevent yawing. This pressure should not be changed to assist the turn.

Use collective pitch control to manage rotor r.p.m. If rotor r.p.m. builds too high during an autorotation, raise the collective sufficiently to decrease r.p.m. back to the normal operating range. If the r.p.m. begins decreasing, you have to again lower the collective. Always keep the rotor r.p.m. within the established range for your helicopter. During a turn, rotor r.p.m. increases due to the increased back cyclic control pressure, which induces a greater airflow through the rotor system. The r.p.m. builds rapidly and can easily exceed the maximum limit if not controlled by use of collective. The tighter the turn and the heavier the gross weight, the higher the r.p.m.

To initiate an autorotation, other than in a low hover, lower the collective pitch control. This holds true whether performing a practice autorotation or in the event of an in-flight engine failure. This reduces the pitch of the main rotor blades and allows them to continue turning at normal r.p.m. During practice autorotations, maintain the r.p.m. in the green arc with the throttle while lowering collective. Once the collective is fully lowered, reduce engine r.p.m. by decreasing the throttle. This causes a split of the engine and rotor r.p.m. needles.

STRAIGHT-IN AUTOROTATION
A straight-in autorotation implies an autorotation from altitude with no turns. The speed at touchdown and the resulting ground run depends on the rate and amount of flare. The greater the degree of flare and the longer it is held, the slower the touchdown speed and the shorter the ground run. The slower the speed desired at touchdown, the more accurate the timing and speed of the flare must be, especially in helicopters with low inertia rotor systems.

TECHNIQUE
Refer to figure 11-1 (position 1). From level flight at the manufacturer's recommended airspeed, between 500 to 700 feet AGL, and heading into the wind, smoothly, but firmly lower the collective pitch control to the full down position, maintaining r.p.m. in the green arc with throttle. Coordinate the collective movement with proper antitorque pedal for trim, and apply aft cyclic control to maintain proper airspeed. Once the collective is fully lowered, decrease throttle to ensure a clean split of the needles. After splitting the needles, readjust the throttle to keep engine r.p.m. above normal idling speed, but not high enough to cause rejoining of the needles. The manufacturer often recommends the proper r.p.m.

At position 2, adjust attitude with cyclic control to obtain the manufacturer's recommended autorotation or best gliding speed. Adjust collective pitch control, as necessary, to maintain rotor r.p.m. in the green arc. Aft cyclic movements cause an increase in rotor r.p.m., which is then controlled by a small increase in collective pitch control. Avoid a large collective pitch increase, which results in a rapid decay of rotor r.p.m.,

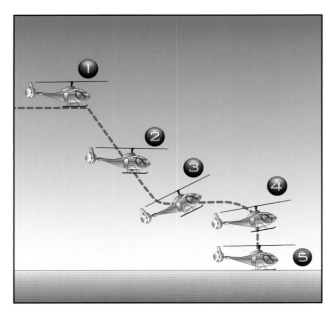

Figure 11-1. Straight-in autorotation.

and leads to "chasing the r.p.m." Avoid looking straight down in front of the aircraft. Continually cross-check attitude, trim, rotor r.p.m., and airspeed.

At approximately 40 to 100 feet above the surface, or at the altitude recommended by the manufacturer (position 3), begin the flare with aft cyclic control to reduce forward airspeed and decrease the rate of descent. Maintain heading with the antitorque pedals. Care must be taken in the execution of the flare so that the cyclic control is not moved rearward so abruptly as to cause the helicopter to climb, nor should it be moved so slowly as to not arrest the descent, which may allow the helicopter to settle so rapidly that the tail rotor strikes the ground. When forward motion decreases to the desired groundspeed, which is usually the slowest possible speed (position 4), move the cyclic control forward to place the helicopter in the proper attitude for landing.

The altitude at this time should be approximately 8 to 15 feet AGL, depending on the altitude recommended by the manufacturer. Extreme caution should be used to avoid an excessive nose high and tail low attitude below 10 feet. At this point, if a full touchdown landing is to be made, allow the helicopter to descend vertically (position 5). Increase collective pitch, as necessary, to check the descent and cushion the landing. Additional antitorque pedal is required to maintain heading as collective pitch is raised due to the reduction in rotor r.p.m. and the resulting reduced effect of the tail rotor. Touch down in a level flight attitude.

A power recovery can be made during training in lieu of a full touchdown landing. Refer to the section on power recoveries for the correct technique.

After touchdown and after the helicopter has come to a complete stop, lower the collective pitch to the full-down position. Do not try to stop the forward ground run with aft cyclic, as the main rotor blades can strike the tail boom. Rather, by lowering the collective slightly during the ground run, more weight is placed on the undercarriage, slowing the helicopter.

COMMON ERRORS

1. Failing to use sufficient antitorque pedal when power is reduced.

2. Lowering the nose too abruptly when power is reduced, thus placing the helicopter in a dive.

3. Failing to maintain proper rotor r.p.m. during the descent.

4. Application of up-collective pitch at an excessive altitude resulting in a hard landing, loss of heading control, and possible damage to the tail rotor and to the main rotor blade stops.

5. Failing to level the helicopter.

POWER RECOVERY FROM PRACTICE AUTOROTATION

A power recovery is used to terminate practice autorotations at a point prior to actual touchdown. After the power recovery, a landing can be made or a go-around initiated.

TECHNIQUE

At approximately 8 to 15 feet above the ground, depending upon the helicopter being used, begin to level the helicopter with forward cyclic control. Avoid excessive nose high, tail low attitude below 10 feet. Just prior to achieving level attitude, with the nose still slightly up, coordinate upward collective pitch control with an increase in the throttle to join the needles at operating r.p.m. The throttle and collective pitch must be coordinated properly. If the throttle is increased too fast or too much, an engine overspeed can occur; if throttle is increased too slowly or too little in proportion to the increase in collective pitch, a loss of rotor r.p.m. results. Use sufficient collective pitch to stop the descent and coordinate proper antitorque pedal pressure to maintain heading. When a landing is to be made following the power recovery, bring the helicopter to a hover at normal hovering altitude and then descend to a landing.

If a go-around is to be made, the cyclic control should be moved forward to resume forward flight. In transitioning from a practice autorotation to a go-around, exercise care to avoid an altitude-airspeed combination that would place the helicopter in an unsafe area of its height-velocity diagram.

COMMON ERRORS

1. Initiating recovery too late, requiring a rapid application of controls, resulting in overcontrolling.

2. Failing to obtain and maintain a level attitude near the surface.

3. Failing to coordinate throttle and collective pitch properly, resulting in either an engine overspeed or a loss of r.p.m.

4. Failing to coordinate proper antitorque pedal with the increase in power

AUTOROTATIONS WITH TURNS

A turn, or a series of turns, can be made during an autorotation in order to land into the wind or avoid obstacles. The turn is usually made early so that the remainder of the autorotation is the same as a straight in autorotation. The most common types are 90° and 180° autorotations. The technique below describes a 180° autorotation.

TECHNIQUE

Establish the aircraft on downwind at recommended airspeed at 700 feet AGL, parallel to the touchdown area. In a no wind or headwind condition, establish the ground track approximately 200 feet away from the touchdown point. If a strong crosswind exists, it will be necessary to move your downwind leg closer or farther out. When abeam the intended touchdown point, reduce collective, and then split the needles. Apply proper antitorque pedal and cyclic to maintain proper attitude. Cross check attitude, trim, rotor r.p.m., and airspeed.

After the descent and airspeed is established, roll into a 180° turn. For training, you should initially roll into a bank of a least 30°, but no more than 40°. Check your airspeed and rotor r.p.m. Throughout the turn, it is important to maintain the proper airspeed and keep the aircraft in trim. Changes in the aircraft's attitude and the angle of bank cause a corresponding change in rotor r.p.m. Adjust the collective, as necessary, in the turn to maintain rotor r.p.m. in the green arc.

At the 90° point, check the progress of your turn by glancing toward your landing area. Plan the second 90 degrees of turn to roll out on the centerline. If you are too close, decrease the bank angle; if too far out, increase the bank angle. Keep the helicopter in trim with antitorque pedals.

The turn should be completed and the helicopter aligned with the intended touchdown area prior to passing through 100 feet AGL. If the collective pitch was increased to control the r.p.m., it may have to be lowered on roll out to prevent a decay in r.p.m. Make an immediate power recovery if the aircraft is not aligned with the touchdown point, and if the rotor r.p.m. and/or airspeed is not within proper limits.

From this point, complete the procedure as if it were a straight-in autorotation.

POWER FAILURE IN A HOVER

Power failures in a hover, also called hovering autorotations, are practiced so that you automatically make the correct response when confronted with engine stoppage or certain other emergencies while hovering. The techniques discussed in this section refer to helicopters with a counter-clockwise rotor system and an antitorque rotor.

TECHNIQUE

To practice hovering autorotations, establish a normal hovering altitude for the particular helicopter being used, considering load and atmospheric conditions. Keep the helicopter headed into the wind and hold maximum allowable r.p.m.

To simulate a power failure, firmly roll the throttle into the spring loaded override position, if applicable. This disengages the driving force of the engine from the rotor, thus eliminating torque effect. As the throttle is closed, apply proper antitorque pedal to maintain heading. Usually, a slight amount of right cyclic control is necessary to keep the helicopter from drifting to the left, to compensate for the loss of tail rotor thrust. However, use cyclic control, as required, to ensure a vertical descent and a level attitude. Leave the collective pitch where it is on entry.

Helicopters with low inertia rotor systems will begin to settle immediately. Keep a level attitude and ensure a vertical descent with cyclic control while maintaining heading with the pedals. At approximately 1 foot above the surface, apply upward collective pitch control, as necessary, to slow the descent and cushion the landing. Usually the full amount of collective pitch is required. As upward collective pitch control is applied, the throttle has to be held in the closed position to prevent the rotor from re-engaging.

Helicopters with high inertia rotor systems will maintain altitude momentarily after the throttle is closed. Then, as the rotor r.p.m. decreases, the helicopter starts to settle. When the helicopter has settled to approximately 1 foot above the surface, apply upward collective pitch control while holding the throttle in the closed position to slow the descent and cushion the landing. The timing of collective pitch control application, and the rate at which it is applied, depends upon the particular helicopter being used, its gross weight, and the existing atmospheric conditions. Cyclic control is used to maintain a level attitude and to ensure a vertical descent. Maintain heading with antitorque pedals.

When the weight of the helicopter is entirely on the skids, cease the application of upward collective. When the helicopter has come to a complete stop, lower the collective pitch to the full down position.

The timing of the collective pitch is a most important consideration. If it is applied too soon, the remaining r.p.m. may not be sufficient to make a soft landing. On the other hand, if collective pitch control is applied too late, surface contact may be made before sufficient blade pitch is available to cushion the landing.

COMMON ERRORS

1. Failing to use sufficient proper antitorque pedal when power is reduced.

2. Failing to stop all sideward or backward movement prior to touchdown.

3. Failing to apply up-collective pitch properly, resulting in a hard touchdown.

4. Failing to touch down in a level attitude.

5. Not rolling the throttle completely to idle.

HEIGHT/VELOCITY DIAGRAM

A height/velocity (H/V) diagram, published by the manufacturer for each model of helicopter, depicts the critical combinations of airspeed and altitude should an engine failure occur. Operating at the altitudes and airspeeds shown within the crosshatched or shaded areas of the H/V diagram may not allow enough time for the critical transition from powered flight to autorotation. [Figure 11-2]

An engine failure in a climb after takeoff occurring in section A of the diagram is most critical. During a climb, a helicopter is operating at higher power settings and blade angle of attack. An engine failure at this point causes a rapid rotor r.p.m. decay because the upward movement of the helicopter must be stopped, then a descent established in order to drive the rotor. Time is also needed to stabilize, then increase the r.p.m. to the normal operating range. The rate of descent must reach a value that is normal for the airspeed at the moment. Since altitude is insufficient for this sequence, you end up with decaying r.p.m., an increasing sink rate, no deceleration lift, little translational lift, and little response to the application of collective pitch to cushion the landing.

It should be noted that, once a steady state autorotation has been established, the H/V diagram no longer applies. An engine failure while descending through section A of the diagram, is less critical, provided a safe landing area is available.

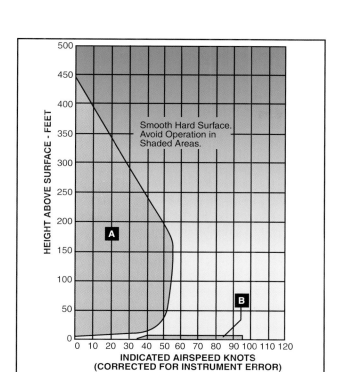

Figure 11-2. By carefully studying the height/velocity diagram, you will be able to avoid the combinations of altitude and airspeed that may not allow you sufficient time or altitude to enter a stabilized autorotative descent. You might want to refer to this diagram during the remainder of the discussion on the height/velocity diagram.

You should avoid the low altitude, high airspeed portion of the diagram (section B), because your recognition of an engine failure will most likely coincide with, or shortly occur after, ground contact. Even if you detect an engine failure, there may not be sufficient time to rotate the helicopter from a nose low, high airspeed attitude to one suitable for slowing, then landing. Additionally, the altitude loss that occurs during recognition of engine failure and rotation to a landing attitude, may not leave enough altitude to prevent the tail skid from hitting the ground during the landing maneuver.

Basically, if the helicopter represented by this H/V diagram is above 445 feet AGL, you have enough time and altitude to enter a steady state autorotation, regardless of your airspeed. If the helicopter is hovering at 5 feet AGL (or less) in normal conditions and the engine fails, a safe hovering autorotation can be made. Between approximately 5 feet and 445 feet AGL, however, the transition to autorotation depends on the altitude and airspeed of the helicopter. Therefore, you should always be familiar with the height/velocity diagram for the particular model of helicopter you are flying.

THE EFFECT OF WEIGHT VERSUS DENSITY ALTITUDE

The height/velocity diagram depicts altitude and airspeed situations from which a successful autorotation

can be made. The time required, and therefore, altitude necessary to attain a steady state autorotative descent, is dependent on the weight of the helicopter and the density altitude. For this reason, the H/V diagram for some helicopter models is valid only when the helicopter is operated in accordance with the gross weight vs. density altitude chart. Where appropriate, this chart is found in the rotorcraft flight manual for the particular helicopter. [Figure 11-3]

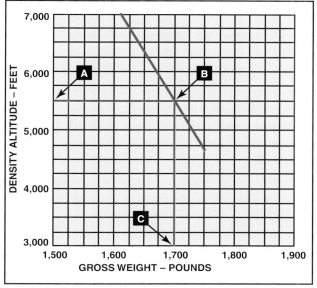

Figure 11-3. Assuming a density altitude of 5,500 feet, the height/velocity diagram in figure 11-2 would be valid up to a gross weight of approximately 1,700 pounds. This is found by entering the graph at a density altitude of 5,500 feet (point A), then moving horizontally to the solid line (point B). Moving vertically to the bottom of the graph (point C), you find that with the existing density altitude, the maximum gross weight under which the height/velocity diagram is applicable is 1,700 pounds.

The gross weight vs. density altitude chart is not intended as a restriction to gross weight, but as an advisory to the autorotative capability of the helicopter during takeoff and climb. You must realize, however, that at gross weights above those recommended by the gross weight vs. density altitude chart, the H/V diagram is not restrictive enough.

VORTEX RING STATE (SETTLING WITH POWER)

Vortex ring state describes an aerodynamic condition where a helicopter may be in a vertical descent with up to maximum power applied, and little or no cyclic authority. The term "settling with power" comes from the fact that helicopter keeps settling even though full engine power is applied.

In a normal out-of-ground-effect hover, the helicopter is able to remain stationary by propelling a large mass of air down through the main rotor. Some of the air is recirculated near the tips of the blades, curling up from the bottom of the rotor system and rejoining the air entering the rotor from the top. This phenomenon is common to all airfoils and is known as tip vortices. Tip vortices consume engine power but produce no useful lift. As long as the tip vortices are small, their only effect is a small loss in rotor efficiency. However, when the helicopter begins to descend vertically, it settles into its own downwash, which greatly enlarges the tip vortices. In this vortex ring state, most of the power developed by the engine is wasted in accelerating the air in a doughnut pattern around the rotor.

In addition, the helicopter may descend at a rate that exceeds the normal downward induced-flow rate of the inner blade sections. As a result, the airflow of the inner blade sections is upward relative to the disc. This produces a secondary vortex ring in addition to the normal tip-vortices. The secondary vortex ring is generated about the point on the blade where the airflow changes from up to down. The result is an unsteady turbulent flow over a large area of the disc. Rotor efficiency is lost even though power is still being supplied from the engine. [Figure 11-4]

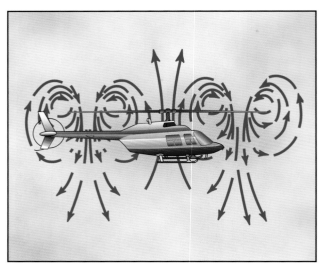

Figure 11-4. Vortex ring state.

A fully developed vortex ring state is characterized by an unstable condition where the helicopter experiences uncommanded pitch and roll oscillations, has little or no cyclic authority, and achieves a descent rate, which, if allowed to develop, may approach 6,000 feet per minute. It is accompanied by increased levels of vibration.

A vortex ring state may be entered during any maneuver that places the main rotor in a condition of high upflow and low forward airspeed. This condition is sometimes seen during quick-stop type maneuvers or during recoveries from autorotations. The following combination of conditions are likely to cause settling in a vortex ring state:

1. A vertical or nearly vertical descent of at least 300 feet per minute. (Actual critical rate depends on the gross weight, r.p.m., density altitude, and other pertinent factors.)

2. The rotor system must be using some of the available engine power (from 20 to 100 percent).

3. The horizontal velocity must be slower than effective translational lift.

Some of the situations that are conducive to a settling with power condition are: attempting to hover out of ground effect at altitudes above the hovering ceiling of the helicopter; attempting to hover out of ground effect without maintaining precise altitude control; or downwind and steep power approaches in which airspeed is permitted to drop to nearly zero.

When recovering from a settling with power condition, the tendency on the part of the pilot is to first try to stop the descent by increasing collective pitch. However, this only results in increasing the stalled area of the rotor, thus increasing the rate of descent. Since inboard portions of the blades are stalled, cyclic control is limited. Recovery is accomplished by increasing forward speed, and/or partially lowering collective pitch. In a fully developed vortex ring state, the only recovery may be to enter autorotation to break the vortex ring state. When cyclic authority is regained, you can then increase forward airspeed.

For settling with power demonstrations and training in recognition of vortex ring state conditions, all maneuvers should be performed at an elevation of at least 1,500 feet AGL.

To enter the maneuver, reduce power below hover power. Hold altitude with aft cyclic until the airspeed approaches 20 knots. Then allow the sink rate to increase to 300 feet per minute or more as the attitude is adjusted to obtain an airspeed of less than 10 knots. When the aircraft begins to shudder, the application of additional up collective increases the vibration and sink rate.

Recovery should be initiated at the first sign of vortex ring state by applying forward cyclic to increase airspeed and simultaneously reducing collective. The recovery is complete when the aircraft passes through effective translational lift and a normal climb is established.

RETREATING BLADE STALL
In forward flight, the relative airflow through the main rotor disc is different on the advancing and retreating side. The relative airflow over the advancing side is higher due to the forward speed of the

helicopter, while the relative airflow on the retreating side is lower. This dissymmetry of lift increases as forward speed increases.

To generate the same amount of lift across the rotor disc, the advancing blade flaps up while the retreating blade flaps down. This causes the angle of attack to decrease on the advancing blade, which reduces lift, and increase on the retreating blade, which increases lift. As the forward speed increases, at some point the low blade speed on the retreating blade, together with its high angle of attack, causes a loss of lift (stall).

Retreating blade stall is a major factor in limiting a helicopter's top forward speed (V_{NE}) and can be felt developing by a low frequency vibration, pitching up of the nose, and a roll in the direction of the retreating blade. High weight, low rotor r.p.m., high density altitude, turbulence and/or steep, abrupt turns are all conducive to retreating blade stall at high forward airspeeds. As altitude is increased, higher blade angles are required to maintain lift at a given airspeed. Thus, retreating blade stall is encountered at a lower forward airspeed at altitude. Most manufacturers publish charts and graphs showing a V_{NE} decrease with altitude.

When recovering from a retreating blade stall condition, moving the cyclic aft only worsens the stall as aft cyclic produces a flare effect, thus increasing angles of attack. Pushing forward on the cyclic also deepens the stall as the angle of attack on the retreating blade is increased. Correct recovery from retreating blade stall requires the collective to be lowered first, which reduces blade angles and thus angle of attack. Aft cyclic can then be used to slow the helicopter.

GROUND RESONANCE

Ground resonance is an aerodynamic phenomenon associated with fully-articulated rotor systems. It develops when the rotor blades move out of phase with each other and cause the rotor disc to become unbalanced. This condition can cause a helicopter to self-destruct in a matter of seconds. However, for this condition to occur, the helicopter must be in contact with the ground.

If you allow your helicopter to touch down firmly on one corner (wheel type landing gear is most conducive for this) the shock is transmitted to the main rotor system. This may cause the blades to move out of their normal relationship with each other. This movement occurs along the drag hinge. [Figure 11-5]

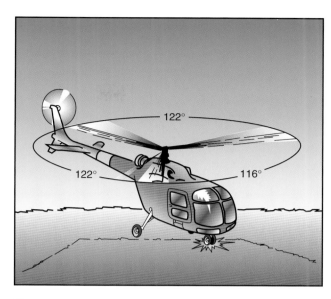

Figure 11-5. Hard contact with the ground can send a shock wave to the main rotor head, resulting in the blades of a three-bladed rotor system moving from their normal 120° relationship to each other. This could result in something like 122°, 122°, and 116° between blades. When one of the other landing gear strikes the surface, the unbalanced condition could be further aggravated.

If the r.p.m. is low, the corrective action to stop ground resonance is to close the throttle immediately and fully lower the collective to place the blades in low pitch. If the r.p.m. is in the normal operating range, you should fly the helicopter off the ground, and allow the blades to automatically realign themselves. You can then make a normal touchdown. If you lift off and allow the helicopter to firmly re-contact the surface before the blades are realigned, a second shock could move the blades again and aggravate the already unbalanced condition. This could lead to a violent, uncontrollable oscillation.

This situation does not occur in rigid or semirigid rotor systems, because there is no drag hinge. In addition, skid type landing gear are not as prone to ground resonance as wheel type gear.

DYNAMIC ROLLOVER

A helicopter is susceptible to a lateral rolling tendency, called dynamic rollover, when lifting off the surface. For dynamic rollover to occur, some factor has to first cause the helicopter to roll or pivot around a skid, or landing gear wheel, until its critical rollover angle is reached. Then, beyond this point, main rotor thrust continues the roll and recovery is impossible. If the critical rollover angle is exceeded, the helicopter rolls on its side regardless of the cyclic corrections made.

Dynamic rollover begins when the helicopter starts to pivot around its skid or wheel. This can occur for a variety of reasons, including the failure to remove a tiedown or skid securing device, or if the skid or wheel contacts a fixed object while hovering sideward, or if the gear is stuck in ice, soft asphalt, or mud. Dynamic rollover may also occur if you do not use the proper landing or takeoff technique or while performing slope operations. Whatever the cause, if the gear or skid becomes a pivot point, dynamic rollover is possible if you do not use the proper corrective technique.

Once started, dynamic rollover cannot be stopped by application of opposite cyclic control alone. For example, the right skid contacts an object and becomes the pivot point while the helicopter starts rolling to the right. Even with full left cyclic applied, the main rotor thrust vector and its moment follows the aircraft as it continues rolling to the right. Quickly applying down collective is the most effective way to stop dynamic rollover from developing. Dynamic rollover can occur in both skid and wheel equipped helicopters, and all types of rotor systems.

CRITICAL CONDITIONS

Certain conditions reduce the critical rollover angle, thus increasing the possibility for dynamic rollover and reducing the chance for recovery. The rate of rolling motion is also a consideration, because as the roll rate increases, the critical rollover angle at which recovery is still possible, is reduced. Other critical conditions include operating at high gross weights with thrust (lift) approximately equal to the weight.

Refer to figure 11-6. The following conditions are most critical for helicopters with counter-clockwise rotor rotation:

1. right side skid/wheel down, since translating tendency adds to the rollover force.

2. right lateral center of gravity.

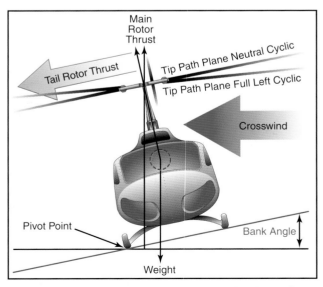

Figure 11-6. Forces acting on a helicopter with right skid on the ground.

3. crosswinds from the left.

4. left yaw inputs.

For helicopters with clockwise rotor rotation, the opposite would be true.

CYCLIC TRIM

When maneuvering with one skid or wheel on the ground, care must be taken to keep the helicopter cyclic control properly trimmed. For example, if a slow takeoff is attempted and the cyclic is not positioned and trimmed to account for translating tendency, the critical recovery angle may be exceeded in less than two seconds. Control can be maintained if you maintain proper cyclic position and trim, and not allow the helicopter's roll and pitch rates to become too great. You should fly your helicopter into the air smoothly while keeping movements of pitch, roll, and yaw small, and not allow any untrimmed cyclic pressures.

NORMAL TAKEOFFS AND LANDINGS

Dynamic rollover is possible even during normal takeoffs and landings on relative level ground, if one wheel or skid is on the ground and thrust (lift) is approximately equal to the weight of the helicopter. If the takeoff or landing is not performed properly, a roll rate could develop around the wheel or skid that is on the ground. When taking off or landing, perform the maneuver smoothly and trim the cyclic so that no pitch or roll movement rates build up, especially the roll rate. If the bank angle starts to increase to an angle of approximately 5 to 8°, and full corrective cyclic does not reduce the angle, the collective should be reduced to diminish the unstable rolling condition.

SLOPE TAKEOFFS AND LANDINGS

During slope operations, excessive application of cyclic control into the slope, together with excessive collective pitch control, can result in the downslope skid rising sufficiently to exceed lateral cyclic control limits, and an upslope rolling motion can occur. [Figure 11-7]

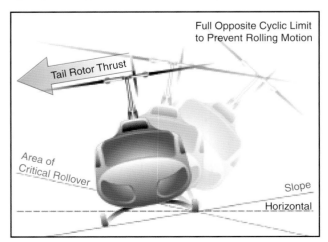

Figure 11-7. Upslope rolling motion.

When performing slope takeoff and landing maneuvers, follow the published procedures and keep the roll rates small. Slowly raise the downslope skid or wheel to bring the helicopter level, and then lift off. During landing, first touch down on the upslope skid or wheel, then slowly lower the downslope skid or wheel using combined movements of cyclic and collective. If the helicopter rolls approximately 5 to 8° to the upslope side, decrease collective to correct the bank angle and return to level attitude, then start the landing procedure again.

USE OF COLLECTIVE

The collective is more effective in controlling the rolling motion than lateral cyclic, because it reduces the main rotor thrust (lift). A smooth, moderate collective reduction, at a rate less than approximately full up to full down in two seconds, is adequate to stop the rolling motion. Take care, however, not to dump collective at too high a rate, as this may cause a main rotor blade to strike the fuselage. Additionally, if the helicopter is on a slope and the roll starts to the upslope side, reducing collective too fast may create a high roll rate in the opposite direction. When the upslope skid/wheel hits the ground, the dynamics of the motion can cause the helicopter to bounce off the upslope skid/wheel, and the inertia can cause the helicopter to roll about the downslope ground contact point and over on its side. [Figure 11-8]

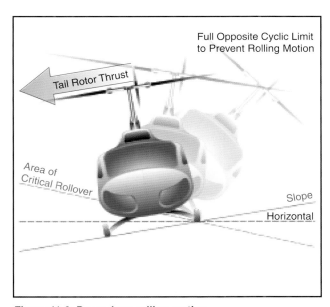

Figure 11-8. Downslope rolling motion.

The collective should not be pulled suddenly to get airborne, as a large and abrupt rolling moment in the opposite direction could occur. Excessive application of collective can result in the upslope skid rising sufficiently to exceed lateral cyclic control limits. This movement may be uncontrollable. If the helicopter develops a roll rate with one skid/wheel on the ground, the helicopter can roll over on its side.

PRECAUTIONS

The following lists several areas to help you avoid dynamic rollover.

1. Always practice hovering autorotations into the wind, but never when the wind is gusty or over 10 knots.

2. When hovering close to fences, sprinklers, bushes, runway/taxi lights, or other obstacles that could catch a skid, use extreme caution.

3. Always use a two-step liftoff. Pull in just enough collective pitch control to be light on the skids and feel for equilibrium, then gently lift the helicopter into the air.

4. When practicing hovering maneuvers close to the ground, make sure you hover high enough to have adequate skid clearance with any obstacles, especially when practicing sideways or rearward flight.

5. When the wind is coming from the upslope direction, less lateral cyclic control will be available.

6. Tailwind conditions should be avoided when conducting slope operations.

7. When the left skid/wheel is upslope, less lateral cyclic control is available due to the translating tendency of the tail rotor. (This is true for counter-rotating rotor systems)

8. If passengers or cargo are loaded or unloaded, the lateral cyclic requirement changes.

9. If the helicopter utilizes interconnecting fuel lines that allow fuel to automatically transfer from one side of the helicopter to the other, the gravitational flow of fuel to the downslope tank could change the center of gravity, resulting in a different amount of cyclic control application to obtain the same lateral result.

10. Do not allow the cyclic limits to be reached. If the cyclic control limit is reached, further lowering of the collective may cause mast bumping. If this occurs, return to a hover and select a landing point with a lesser degree of slope.

11. During a takeoff from a slope, if the upslope skid/wheel starts to leave the ground before the downslope skid/wheel, smoothly and gently

lower the collective and check to see if the downslope skid/wheel is caught on something. Under these conditions vertical ascent is the only acceptable method of liftoff.

12. During flight operations on a floating platform, if the platform is pitching/rolling while attempting to land or takeoff, the result could be dynamic rollover.

LOW G CONDITIONS AND MAST BUMPING

For cyclic control, small helicopters depend primarily on tilting the main rotor thrust vector to produce control moments about the aircraft center of gravity (CG), causing the helicopter to roll or pitch in the desired direction. Pushing the cyclic control forward abruptly from either straight-and-level flight or after a climb can put the helicopter into a low G (weightless) flight condition. In forward flight, when a push-over is performed, the angle of attack and thrust of the rotor is reduced, causing a low G or weightless flight condition. During the low G condition, the lateral cyclic has little, if any, effect because the rotor thrust has been reduced. Also, in a counter-clockwise rotor system (a clockwise system would be the reverse), there is no main rotor thrust component to the left to counteract the tail rotor thrust to the right, and since the tail rotor is above the CG, the tail rotor thrust causes the helicopter to roll rapidly to the right. If you attempt to stop the right roll by applying full left cyclic before regaining main rotor thrust, the rotor can exceed its flapping limits and cause structural failure of the rotor shaft due to mast bumping, or it may allow a blade to contact the airframe. [Figure 11-9]

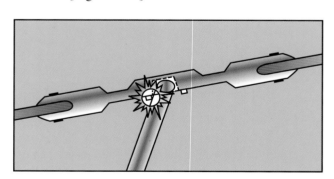

Figure 11-9. In a low G condition, improper corrective action could lead to the main rotor hub contacting the rotor mast. The contact with the mast becomes more violent with each successive flapping motion. This, in turn, creates a greater flapping displacement. The result could be a severely damaged rotor mast, or the main rotor system could separate from the helicopter.

Since a low G condition could have disastrous results, the best way to prevent it from happening is to avoid the conditions where it might occur. This means avoiding turbulence as much as possible. If you do encounter turbulence, slow your forward airspeed and make small control inputs. If turbulence becomes excessive, consider making a precautionary landing. To help prevent turbulence induced inputs, make sure your cyclic arm is properly supported. One way to accomplish this is to brace your arm against your leg. Even if you are not in turbulent conditions, you should avoid abrupt movement of the cyclic and collective.

If you do find yourself in a low G condition, which can be recognized by a feeling of weightlessness and an uncontrolled roll to the right, you should immediately and smoothly apply aft cyclic. Do not attempt to correct the rolling action with lateral cyclic. By applying aft cyclic, you will load the rotor system, which in turn produces thrust. Once thrust is restored, left cyclic control becomes effective, and you can roll the helicopter to a level attitude.

LOW ROTOR RPM AND BLADE STALL

As mentioned earlier, low rotor r.p.m. during an autorotation might result in a less than successful maneuver. However, if you let rotor r.p.m. decay to the point where all the rotor blades stall, the result is usually fatal, especially when it occurs at altitude. The danger of low rotor r.p.m. and blade stall is greatest in small helicopters with low blade inertia. It can occur in a number of ways, such as simply rolling the throttle the wrong way, pulling more collective pitch than power available, or when operating at a high density altitude.

When the rotor r.p.m. drops, the blades try to maintain the same amount of lift by increasing pitch. As the pitch increases, drag increases, which requires more power to keep the blades turning at the proper r.p.m. When power is no longer available to maintain r.p.m., and therefore lift, the helicopter begins to descend. This changes the relative wind and further increases the angle of attack. At some point the blades will stall unless r.p.m. is restored. If all blades stall, it is almost impossible to get smooth air flowing across the blades.

Even though there is a safety factor built into most helicopters, anytime your rotor r.p.m. falls below the green arc, and you have power, simultaneously add throttle and lower the collective. If you are in forward flight, gently applying aft cyclic loads up the rotor system and helps increase rotor r.p.m. If you are without power, immediately lower the collective and apply aft cyclic.

RECOVERY FROM LOW ROTOR RPM

Under certain conditions of high weight, high temperature, or high density altitude, you might get into a situation where the r.p.m. is low even though you are using maximum throttle. This is usually the result of the main rotor blades having an angle of attack that has created so much drag that engine power is not sufficient to maintain or attain normal operating r.p.m.

If you are in a low r.p.m. situation, the lifting power of the main rotor blades can be greatly diminished. As soon as you detect a low r.p.m. condition, immediately apply additional throttle, if available, while slightly lowering the collective. This reduces main rotor pitch and drag. As the helicopter begins to settle, smoothly raise the collective to stop the descent. At hovering altitude you may have to repeat this technique several times to regain normal operating r.p.m. This technique is sometimes called "milking the collective." When operating at altitude, the collective may have to be lowered only once to regain rotor speed. The amount the collective can be lowered depends on altitude. When hovering near the surface, make sure the helicopter does not contact the ground as the collective is lowered.

Since the tail rotor is geared to the main rotor, low main rotor r.p.m. may prevent the tail rotor from producing enough thrust to maintain directional control. If pedal control is lost and the altitude is low enough that a landing can be accomplished before the turning rate increases dangerously, slowly decrease collective pitch, maintain a level attitude with cyclic control, and land.

SYSTEM MALFUNCTIONS

The reliability and dependability record of modern helicopters is very impressive. By following the manufacturer's recommendations regarding periodic maintenance and inspections, you can eliminate most systems and equipment failures. Most malfunctions or failures can be traced to some error on the part of the pilot; therefore, most emergencies can be averted before they happen. An actual emergency is a rare occurrence.

ANTITORQUE SYSTEM FAILURE

Antitorque failures usually fall into two categories. One focuses on failure of the power drive portion of the tail rotor system resulting in a complete loss of antitorque. The other category covers mechanical control failures where the pilot is unable to change or control tail rotor thrust even though the tail rotor may still be providing antitorque thrust.

Tail rotor drive system failures include driveshaft failures, tail rotor gearbox failures, or a complete loss of the tail rotor itself. In any of these cases, the loss of antitorque normally results in an immediate yawing of the helicopter's nose. The helicopter yaws to the right in a counter-clockwise rotor system and to the left in a clockwise system. This discussion assumes a helicopter with a counter-clockwise rotor system. The severity of the yaw is proportionate to the amount of power being used and the airspeed. An antitorque failure with a high power setting at a low airspeed

results in a severe yawing to the right. At low power settings and high airspeeds, the yaw is less severe. High airspeeds tend to streamline the helicopter and keep it from spinning.

If a tail rotor failure occurs, power has to be reduced in order to reduce main rotor torque. The techniques differ depending on whether the helicopter is in flight or in a hover, but will ultimately require an autorotation. If a complete tail rotor failure occurs while hovering, enter a hovering autorotation by rolling off the throttle. If the failure occurs in forward flight, enter a normal autorotation by lowering the collective and rolling off the throttle. If the helicopter has enough forward airspeed (close to cruising speed) when the failure occurs, and depending on the helicopter design, the vertical stabilizer may provide enough directional control to allow you to maneuver the helicopter to a more desirable landing sight. Some of the yaw may be compensated for by applying slight cyclic control opposite the direction of yaw. This helps in directional control, but also increases drag. Care must be taken not to lose too much forward airspeed because the streamlining effect diminishes as airspeed is reduced. Also, more altitude is required to accelerate to the correct airspeed if an autorotation is entered into at a low airspeed.

A mechanical control failure limits or prevents control of tail rotor thrust and is usually caused by a stuck or broken control rod or cable. While the tail rotor is still producing antitorque thrust, it cannot be controlled by the pilot. The amount of antitorque depends on the position where the controls jam or fail. Once again, the techniques differ depending on the amount of tail rotor thrust, but an autorotation is generally not required.

LANDING—STUCK LEFT PEDAL

Be sure to follow the procedures and techniques outlined in the FAA-approved rotorcraft flight manual for the helicopter you are flying. A stuck left pedal, such as might be experienced during takeoff or climb conditions, results in the helicopter's nose yawing to the left when power is reduced. Rolling off the throttle and entering an autorotation only makes matters worse. The landing profile for a stuck left pedal is best described as a normal approach to a momentary hover at three to four feet above the surface. Following an analysis, make the landing. If the helicopter is not turning, simply lower the helicopter to the surface. If the helicopter is turning to the right, roll the throttle toward flight idle the amount necessary to stop the turn as you land. If the helicopter is beginning to turn left, you should be able to make the landing prior to the turn rate becoming excessive. However, if the turn rate becomes excessive prior to the landing, simply execute a takeoff and return for another landing.

LANDING—STUCK NEUTRAL OR RIGHT PEDAL

The landing profile for a stuck neutral or a stuck right pedal is a low power approach or descent with a running or roll-on landing. The approach profile can best be described as a steep approach with a flare at the bottom to slow the helicopter. The power should be low enough to establish a left yaw during the descent. The left yaw allows a margin of safety due to the fact that the helicopter will turn to the right when power is applied. This allows the momentary use of power at the bottom of the approach. As you apply power, the helicopter rotates to the right and becomes aligned with the landing area. At this point, roll the throttle to flight idle and make the landing. The momentary use of power helps stop the descent and allows additional time for you to level the helicopter prior to closing the throttle.

If the helicopter is not yawed to the left at the conclusion of the flare, roll the throttle to flight idle and use the collective to cushion the touchdown. As with any running or roll-on landing, use the cyclic to maintain the ground track. This technique results in a longer ground run or roll than if the helicopter was yawed to the left.

UNANTICIPATED YAW / LOSS OF TAIL ROTOR EFFECTIVENESS (LTE)

Unanticipated yaw is the occurrence of an uncommanded yaw rate that does not subside of its own accord and, which, if not corrected, can result in the loss of helicopter control. This uncommanded yaw rate is referred to as loss of tail rotor effectiveness (LTE) and occurs to the right in helicopters with a counter-clockwise rotating main rotor and to the left in helicopters with a clockwise main rotor rotation. Again, this discussion covers a helicopter with a counter-clockwise rotor system and an antitorque rotor.

LTE is not related to an equipment or maintenance malfunction and may occur in all single-rotor helicopters at airspeeds less than 30 knots. It is the result of the tail rotor not providing adequate thrust to maintain directional control, and is usually caused by either certain wind azimuths (directions) while hovering, or by an insufficient tail rotor thrust for a given power setting at higher altitudes.

For any given main rotor torque setting in perfectly steady air, there is an exact amount of tail rotor thrust required to prevent the helicopter from yawing either left or right. This is known as tail rotor trim thrust. In order to maintain a constant heading while hovering, you should maintain tail rotor thrust equal to trim thrust.

The required tail rotor thrust is modified by the effects of the wind. The wind can cause an uncommanded yaw by changing tail rotor effective thrust. Certain relative wind directions are more likely to cause tail rotor thrust variations than others. Flight and wind tunnel tests have identified three relative wind azimuth regions that can either singularly, or in combination, create an LTE conducive environment. These regions can overlap, and thrust variations may be more pronounced. Also, flight testing has determined that the tail rotor does not actually stall during the period. When operating in these areas at less than 30 knots, pilot workload increases dramatically.

MAIN ROTOR DISC INTERFERENCE (285-315°)

Refer to figure 11-10. Winds at velocities of 10 to 30 knots from the left front cause the main rotor vortex to be blown into the tail rotor by the relative wind. The effect of this main rotor disc vortex causes the tail rotor to operated in an extremely turbulent environment. During a right turn, the tail rotor experiences a reduction of thrust as it comes into the area of the main rotor disc vortex. The reduction in tail rotor thrust comes from the airflow changes experienced at the tail rotor as the main rotor disc vortex moves across the tail rotor disc. The effect of the main rotor disc vortex initially increases the angle of attack of the tail rotor blades, thus increasing tail rotor thrust. The increase in the angle of attack requires that right pedal pressure be added to reduce tail rotor thrust in order to maintain the same rate of turn. As the main rotor vortex passes the tail rotor, the tail rotor angle of attack is reduced. The reduction in the angle of attack causes a reduction in thrust and a right yaw acceleration begins. This acceleration can be surprising, since you were previously adding right pedal to maintain the right turn rate. This thrust reduction occurs suddenly, and if uncorrected, develops into an uncontrollable rapid rotation about the mast. When operating within this region, be aware that the reduction in tail rotor thrust can happen quite suddenly, and be prepared to react quickly to counter this reduction with additional left pedal input.

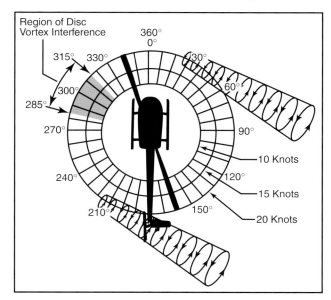

Figure 11-10. Main rotor disc vortex interference.

WEATHERCOCK STABILITY (120-240°)

In this region, the helicopter attempts to weathervane its nose into the relative wind. [Figure 11-11] Unless a resisting pedal input is made, the helicopter starts a slow, uncommanded turn either to the right or left depending upon the wind direction. If the pilot allows a right yaw rate to develop and the tail of the helicopter moves into this region, the yaw rate can accelerate rapidly. In order to avoid the onset of LTE in this downwind condition, it is imperative to maintain positive control of the yaw rate and devote full attention to flying the helicopter.

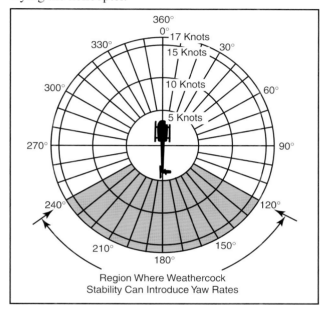

Figure 11-11. Weathercock stability.

TAIL ROTOR VORTEX RING STATE (210-330°)

Winds within this region cause a tail rotor vortex ring state to develop. [Figure 11-12] The result is a non-uniform, unsteady flow into the tail rotor. The vortex ring state causes tail rotor thrust variations, which result in yaw deviations. The net effect of the unsteady flow is an oscillation of tail rotor thrust. Rapid and continuous pedal movements are necessary to compensate for the rapid changes in tail rotor thrust when hovering in a left crosswind. Maintaining a precise heading in this region is difficult, but this characteristic presents no significant problem unless corrective action is delayed. However, high pedal workload, lack of concentration and overcontrolling can all lead to LTE.

When the tail rotor thrust being generated is less than the thrust required, the helicopter yaws to the right. When hovering in left crosswinds, you must concentrated on smooth pedal coordination and not allow an uncontrolled right yaw to develop. If a right yaw rate is allowed to build, the helicopter can rotate into the wind azimuth region where weathercock stability then accelerates the right turn rate. Pilot workload during a tail rotor vortex ring state is high. Do not allow a right yaw rate to increase.

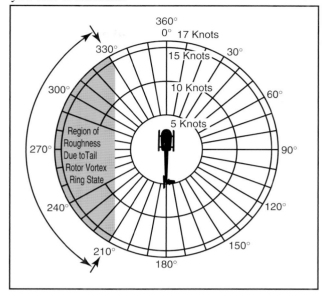

Figure 11-12. Tail rotor vortex ring state.

LTE AT ALTITUDE

At higher altitudes, where the air is thinner, tail rotor thrust and efficiency is reduced. When operating at high altitudes and high gross weights, especially while hovering, the tail rotor thrust may not be sufficient to maintain directional control and LTE can occur. In this case, the hovering ceiling is limited by tail rotor thrust and not necessarily power available. In these conditions gross weights need to be reduced and/or operations need to be limited to lower density altitudes.

REDUCING THE ONSET OF LTE

To help reduce the onset of loss of tail rotor effectiveness, there are some steps you can follow.

1. Maintain maximum power-on rotor r.p.m. If the main rotor r.p.m. is allowed to decrease, the antitorque thrust available is decreased proportionally.

2. Avoid tailwinds below an airspeed of 30 knots. If loss of translational lift occurs, it results in an increased power demand and additional antitorque pressures.

3. Avoid out of ground effect (OGE) operations and high power demand situations below an airspeed of 30 knots.

4. Be especially aware of wind direction and velocity when hovering in winds of about 8-12 knots. There are no strong indicators that translational lift has been reduced. A loss of translational lift results in an unexpected high power demand and an increased antitorque requirement.

5. Be aware that if a considerable amount of left pedal is being maintained, a sufficient amount of left pedal may not be available to counteract an unanticipated right yaw.

6. Be alert to changing wind conditions, which may be experienced when flying along ridge lines and around buildings.

RECOVERY TECHNIQUE

If a sudden unanticipated right yaw occurs, the following recovery technique should be performed. Apply full left pedal while simultaneously moving cyclic control forward to increase speed. If altitude permits, reduce power. As recovery is effected, adjust controls for normal forward flight.

Collective pitch reduction aids in arresting the yaw rate but may cause an excessive rate of descent. Any large, rapid increase in collective to prevent ground or obstacle contact may further increase the yaw rate and decrease rotor r.p.m. The decision to reduce collective must be based on your assessment of the altitude available for recovery.

If the rotation cannot be stopped and ground contact is imminent, an autorotation may be the best course of action. Maintain full left pedal until the rotation stops, then adjust to maintain heading.

MAIN DRIVE SHAFT FAILURE

The main drive shaft, located between the engine and the main rotor gearbox, transmits engine power to the main rotor gearbox. In some helicopters, particularly those with piston engines, a drive belt is used instead of a drive shaft. A failure of the drive shaft or belt has the same effect as an engine failure, because power is no longer provided to the main rotor, and an autorotation has to be initiated. There are a few differences, however, that need to be taken into consideration. If the drive shaft or belt breaks, the lack of any load on the engine results in an overspeed. In this case, the throttle must be closed in order to prevent any further damage. In some helicopters, the tail rotor drive system continues to be powered by the engine even if the main drive shaft breaks. In this case, when the engine unloads, a tail rotor overspeed can result. If this happens, close the throttle immediately and enter an autorotation.

HYDRAULIC FAILURES

Most helicopters, other than smaller piston powered helicopters, incorporate the use of hydraulic actuators to overcome high control forces. A hydraulic system consists of actuators, also called servos, on each flight control; a pump, which is usually driven by the main rotor gearbox; and a reservoir to store the hydraulic fluid. A switch in the cockpit can turn the system off, although it is left on under normal conditions. A pressure indicator in the cockpit may be installed to monitor the system.

An impending hydraulic failure can be recognized by a grinding or howling noise from the pump or actuators, increased control forces and feedback, and limited control movement. The corrective action required is stated in detail in the appropriate rotorcraft flight manual. However, in most cases, airspeed needs to be reduced in order to reduce control forces. The hydraulic switch and circuit breaker should be checked and recycled. If hydraulic power is not restored, make a shallow approach to a running or roll-on landing. This technique is used because it requires less control force and pilot workload. Additionally, the hydraulic system should be disabled, by either pulling the circuit breaker and/or placing the switch in the off position. The reason for this is to prevent an inadvertent restoration of hydraulic power, which may lead to overcontrolling near the ground.

In those helicopters where the control forces are so high that they cannot be moved without hydraulic assistance, two or more independent hydraulic systems may be installed. Some helicopters use hydraulic accumulators to store pressure that can be used for a short time while in an emergency if the hydraulic pump fails. This gives you enough time to land the helicopter with normal control.

GOVERNOR FAILURE

Governors automatically adjust engine power to maintain rotor r.p.m. when the collective pitch is changed. If the governor fails, any change in collective pitch requires you to manually adjust the throttle to maintain correct r.p.m. In the event of a high side governor failure, the engine and rotor r.p.m. try to increase above the normal range. If the r.p.m. cannot be reduced and controlled with the throttle, close the throttle and enter an autorotation. If the governor fails on the low side, normal r.p.m. may not be attainable, even if the throttle is manually controlled. In this case, the collective has to be lowered to maintain r.p.m. A running or roll-on landing may be performed if the engine can maintain sufficient rotor r.p.m. If there is insufficient power, enter an autorotation.

ABNORMAL VIBRATIONS

With the many rotating parts found in helicopters, some vibration is inherent. You need to understand the cause and effect of helicopter vibrations because abnormal vibrations cause premature component wear and may even result in structural failure. With experience, you learn what vibrations are normal versus those that are abnormal and can then decide whether continued flight is safe or not. Helicopter vibrations are categorized into low, medium, or high frequency.

LOW FREQUENCY VIBRATIONS

Low frequency vibrations (100-500 cycles per minute) usually originate from the main rotor system. The vibration may be felt through the controls, the airframe, or a combination of both. Furthermore, the vibration may have a definite direction of push or thrust. It may be vertical, lateral, horizontal, or even a combination. Normally, the direction of the vibration can be determined by concentrating on the feel of the vibration, which may push you up and down, backwards and forwards, or from side to side. The direction of the vibration and whether it is felt in the controls or the airframe is an important means for the mechanic to troubleshoot the source. Some possible causes could be that the main rotor blades are out of track or balance, damaged blades, worn bearings, dampers out of adjustment, or worn parts.

MEDIUM AND HIGH FREQUENCY VIBRATIONS

Medium frequency vibrations (1,000 - 2,000 cycles per minute) and high frequency vibrations (2,000 cycles per minute or higher) are normally associated with out-of-balance components that rotate at a high r.p.m., such as the tail rotor, engine, cooling fans, and components of the drive train, including transmissions, drive shafts, bearings, pulleys, and belts. Most tail rotor vibrations can be felt through the tail rotor pedals as long as there are no hydraulic actuators, which usually dampen out the vibration. Any imbalance in the tail rotor system is very harmful, as it can cause cracks to develop and rivets to work loose. Piston engines usually produce a normal amount of high frequency vibration, which is aggravated by engine malfunctions such as spark plug fouling, incorrect magneto timing, carburetor icing and/or incorrect fuel/air mixture. Vibrations in turbine engines are often difficult to detect as these engines operate at a very high r.p.m.

TRACKING AND BALANCE

Modern equipment used for tracking and balancing the main and tail rotor blades can also be used to detect other vibrations in the helicopter. These systems use accelerometers mounted around the helicopter to detect the direction, frequency, and intensity of the vibration. The built-in software can then analyze the information, pinpoint the origin of the vibration, and suggest the corrective action.

FLIGHT DIVERSION

There will probably come a time in your flight career when you will not be able to make it to your destination. This can be the result of unpredictable weather conditions, a system malfunction, or poor preflight planning. In any case, you will need to be able to safely and efficiently divert to an alternate destination. Before any cross-country flight, check the charts for airports or suitable landing areas along or near your route of flight. Also, check for navaids that can be used during a diversion.

Computing course, time, speed, and distance information in flight requires the same computations used during preflight planning. However, because of the limited cockpit space, and because you must divide your attention between flying the helicopter, making calculations, and scanning for other aircraft, you should take advantage of all possible shortcuts and rule-of-thumb computations.

When in flight, it is rarely practical to actually plot a course on a sectional chart and mark checkpoints and distances. Furthermore, because an alternate airport is usually not very far from your original course, actual plotting is seldom necessary.

A course to an alternate can be measured accurately with a protractor or plotter, but can also be measured with reasonable accuracy using a straightedge and the compass rose depicted around VOR stations. This approximation can be made on the basis of a radial from a nearby VOR or an airway that closely parallels the course to your alternate. However, you must remember that the magnetic heading associated with a VOR radial or printed airway is outbound from the station. To find the course TO the station, it may be necessary to determine the reciprocal of the indicated heading.

Distances can be determined by using a plotter, or by placing a finger or piece of paper between the two and then measuring the approximate distance on the mileage scale at the bottom of the chart.

Before changing course to proceed to an alternate, you should first consider the relative distance and route of flight to all suitable alternates. In addition, you should consider the type of terrain along the route. If circumstances warrant, and your helicopter is equipped with navigational equipment, it is typically easier to navigate to an alternate airport that has a VOR or NDB facility on the field.

After you select the most appropriate alternate, approximate the magnetic course to the alternate using a compass rose or airway on the sectional chart. If time permits, try to start the diversion over a prominent ground feature. However, in an emergency, divert promptly toward your alternate. To complete all plotting, measuring, and computations involved before diverting to the alternate may only aggravate an actual emergency.

Once established on course, note the time, and then use the winds aloft nearest to your diversion point to calculate a heading and groundspeed. Once you have calculated your groundspeed, determine a new arrival time and fuel consumption.

You must give priority to flying the helicopter while dividing your attention between navigation and planning. When determining an altitude to use while diverting, you should consider cloud heights, winds, terrain, and radio reception.

LOST PROCEDURES

Getting lost in an aircraft is a potentially dangerous situation especially when low on fuel. Helicopters have an advantage over airplanes, as they can land almost anywhere before they run out of fuel.

If you are lost, there are some good common sense procedures to follow. If you are nowhere near or cannot see a town or city, the first thing you should do is climb. An increase in altitude increases radio and navigation reception range, and also increases radar coverage. If you are flying near a town or city, you may be able to read the name of the town on a water tower or even land to ask directions.

If your helicopter has a navigational radio, such as a VOR or ADF receiver, you can possibly determine your position by plotting your azimuth from two or more navigational facilities. If GPS is installed, or you have a portable aviation GPS on board, you can use it to determine your position and the location of the nearest airport.

Communicate with any available facility using frequencies shown on the sectional chart. If you are able to communicate with a controller, you may be offered radar vectors. Other facilities may offer direction finding (DF) assistance. To use this procedure, the controller will request you to hold down your transmit button for a few seconds and then release it. The controller may ask you to change directions a few times and repeat the transmit procedure. This gives the controller enough information to plot your position and then give you vectors to a suitable landing sight. If your situation becomes threatening, you can transmit your problems on the emergency frequency 121.5 MHZ and set your transponder to 7700. Most facilities, and even airliners, monitor the emergency frequency.

EMERGENCY EQUIPMENT AND SURVIVAL GEAR

Both Canada and Alaska require pilots to carry survival gear. However, it is good common sense that any time you are flying over rugged and desolated terrain, consider carrying survival gear. Depending on the size and storage capacity of your helicopter, the following are some suggested items:

- Food that is not subject to deterioration due to heat or cold. There should be at least 10,000 calo-

ries for each person on board, and it should be stored in a sealed waterproof container. It should have been inspected by the pilot or his representative within the previous six months, and bear a label verifying the amount and satisfactory condition of the contents.

- A supply of water.

- Cooking utensils.

- Matches in a waterproof container.

- A portable compass.

- An ax at least 2.5 pounds with a handle not less than 28 inches in length.

- A flexible saw blade or equivalent cutting tool.

- 30 feet of snare wire and instructions for use.

- Fishing equipment, including still-fishing bait and gill net with not more than a two inch mesh.

- Mosquito nets or netting and insect repellent sufficient to meet the needs of all persons aboard, when operating in areas where insects are likely to be hazardous.

- A signaling mirror.

- At least three pyrotechnic distress signals.

- A sharp, quality jackknife or hunting knife.

- A suitable survival instruction manual.

- Flashlight with spare bulbs and batteries.

- Portable ELT with spare batteries.

Additional items when there are no trees:

- Stove with fuel or a self-contained means of providing heat for cooking.

- Tent(s) to accommodate everyone on board.

Additional items for winter operations:

- Winter sleeping bags for all persons when the temperature is expected to be below 7°C.

- Two pairs of snow shoes.

- Spare ax handle.

- Honing stone or file.

- Ice chisel.

- Snow knife or saw knife.

Attitude instrument flying in helicopters is essentially visual flying with the flight instruments substituted for the various reference points on the helicopter and the natural horizon. Control changes, required to produce a given attitude by reference to instruments, are identical to those used in helicopter VFR flight, and your thought processes are the same. Basic instrument training is intended as a building block towards attaining an instrument rating. It will also enable you to do a 180° turn in case of inadvertent incursion into instrument meteorological conditions (IMC).

FLIGHT INSTRUMENTS

When flying a helicopter with reference to the flight instruments, proper instrument interpretation is the basis for aircraft control. Your skill, in part, depends on your understanding of how a particular instrument or system functions, including its indications and limitations. With this knowledge, you can quickly determine what an instrument is telling you and translate that information into a control response.

PITOT-STATIC INSTRUMENTS

The pitot-static instruments, which include the airspeed indicator, altimeter, and vertical speed indicator, operate on the principle of differential air pressure. Pitot pressure, also called impact, ram, or dynamic pressure, is directed only to the airspeed indicator, while static pressure, or ambient pressure, is directed to all three instruments. An alternate static source may be included allowing you to select an alternate source of ambient pressure in the event the main port becomes blocked. [Figure 12-1]

AIRSPEED INDICATOR

The airspeed indicator displays the speed of the helicopter through the air by comparing ram air pressure from the pitot tube with static air pressure from the static port—the greater the differential, the greater the speed. The instrument displays the result of this pressure differential as indicated airspeed (IAS). Manufacturers use this speed as the basis for determining helicopter performance, and it may be displayed in knots, miles per hour, or both. [Figure 12-2] When an indicated airspeed is given for a particular situation, you normally use that speed without making a correction for altitude or temperature. The reason no correction is needed is that an airspeed indicator and aircraft performance are affected equally by changes in air density. An indicated airspeed always yields the same performance because the indicator has, in fact, compensated for the change in the environment.

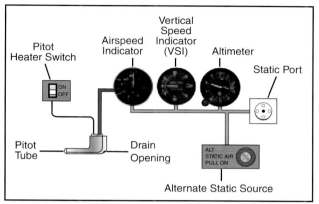

Figure 12-1. Ram air pressure is supplied only to the airspeed indicator, while static pressure is used by all three instruments. Electrical heating elements may be installed to prevent ice from forming on the pitot tube. A drain opening to remove moisture is normally included.

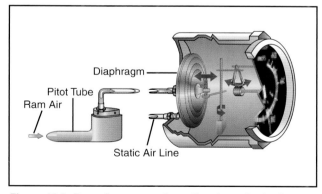

Figure 12-2. Ram air pressure from the pitot tube is directed to a diaphragm inside the airspeed indicator. The airtight case is vented to the static port. As the diaphragm expands or contracts, a mechanical linkage moves the needle on the face of the indicator.

INSTRUMENT CHECK—During the preflight, ensure that the pitot tube, drain hole, and static ports are unobstructed. Before liftoff, make sure the airspeed indicator is reading zero. If there is a strong wind blowing directly at the helicopter, the airspeed indicator may read higher than zero, depending on the wind speed and direction. As you begin your takeoff, make sure the airspeed indicator is increasing at an appropriate rate. Keep in mind, however, that the airspeed indication might be unreliable below a certain airspeed due to rotor downwash.

ALTIMETER

The altimeter displays altitude in feet by sensing pressure changes in the atmosphere. There is an adjustable barometric scale to compensate for changes in atmospheric pressure. [Figure 12-3]

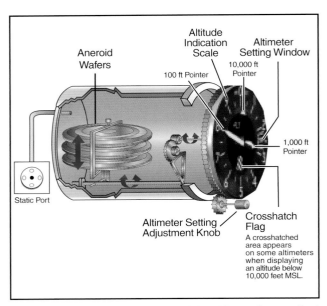

Figure 12-3. The main component of the altimeter is a stack of sealed aneroid wafers. They expand and contract as atmospheric pressure from the static source changes. The mechanical linkage translates these changes into pointer movements on the indicator.

The basis for altimeter calibration is the International Standard Atmosphere (ISA), where pressure, temperature, and lapse rates have standard values. However, actual atmospheric conditions seldom match the standard values. In addition, local pressure readings within a given area normally change over a period of time, and pressure frequently changes as you fly from one area to another. As a result, altimeter indications are subject to errors, the extent of which depends on how much the pressure, temperature, and lapse rates deviate from standard, as well as how recently you have set the altimeter. The best way to minimize altimeter errors is to update the altimeter setting frequently. In most cases, use the current altimeter setting of the nearest reporting station along your route of flight per regulatory requirements.

INSTRUMENT CHECK—During the preflight, ensure that the static ports are unobstructed. Before lift-off, set the altimeter to the current setting. If the altimeter indicates within 75 feet of the actual elevation, the altimeter is generally considered acceptable for use.

VERTICAL SPEED INDICATOR

The vertical speed indicator (VSI) displays the rate of climb or descent in feet per minute (f.p.m.) by measuring how fast the ambient air pressure increases or decreases as the helicopter changes altitude. Since the VSI measures only the rate at which air pressure changes, air temperature has no effect on this instrument. [Figure 12-4]

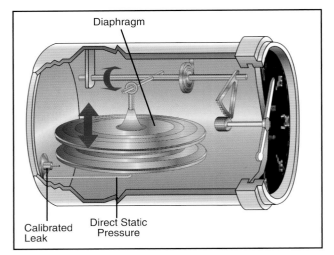

Figure 12-4. Although the sealed case and diaphragm are both connected to the static port, the air inside the case is restricted through a calibrated leak. When the pressures are equal, the needle reads zero. As you climb or descend, the pressure inside the diaphragm instantly changes, and the needle registers a change in vertical direction. When the pressure differential stabilizes at a definite ratio, the needle registers the rate of altitude change.

There is a lag associated with the reading on the VSI, and it may take a few seconds to stabilize when showing rate of climb or descent. Rough control technique and turbulence can further extend the lag period and cause erratic and unstable rate indications. Some aircraft are equipped with an instantaneous vertical speed indicator (IVSI), which incorporates accelerometers to compensate for the lag found in the typical VSI.

INSTRUMENT CHECK—During the preflight, ensure that the static ports are unobstructed. Check to see that the VSI is indicating zero before lift-off. During takeoff, check for a positive rate of climb indication.

SYSTEM ERRORS

The pitot-static system and associated instruments are usually very reliable. Errors are generally caused when the pitot or static openings are blocked. This may be caused by dirt, ice formation, or insects. Check the pitot and static openings for obstructions during the preflight. It is also advisable to place covers on the pitot and static ports when the helicopter is parked on the ground.

The airspeed indicator is the only instrument affected by a blocked pitot tube. The system can become clogged in two ways. If the ram air inlet is clogged, but the drain hole remains open, the airspeed indicator registers zero, regardless of airspeed. If both the ram air inlet and the drain hole become blocked, pressure in the line is trapped, and the airspeed indicator reacts like an altimeter, showing an increase in airspeed with an increase in altitude, and a decrease in speed as altitude decreases. This occurs as long as the static port remains unobstructed.

If the static port alone becomes blocked, the airspeed indicator continues to function, but with incorrect readings. When you are operating above the altitude where the static port became clogged, the airspeed indicator reads lower than it should. Conversely, when operating below that altitude, the indicator reads higher than the correct value. The amount of error is proportional to the distance from the altitude where the static system became blocked. The greater the difference, the greater the error. With a blocked static system, the altimeter freezes at the last altitude and the VSI freezes at zero. Both instruments are then unusable.

Some helicopters are equipped with an alternate static source, which may be selected in the event that the main static system becomes blocked. The alternate source generally vents into the cabin, where air pressures are slightly different than outside pressures, so the airspeed and altimeter usually read higher than normal. Correction charts may be supplied in the flight manual.

GYROSCOPIC INSTRUMENTS
The three gyroscopic instruments that are required for instrument flight are the attitude indicator, heading indicator, and turn indicator. When installed in helicopters, these instruments are usually electrically powered.

Gyros are affected by two principles—rigidity in space and precession. Rigidity in space means that once a gyro is spinning, it tends to remain in a fixed position and resists external forces applied to it. This principle allows a gyro to be used to measure changes in attitude or direction.

Precession is the tilting or turning of a gyro in response to pressure. The reaction to this pressure does not occur at the point where it was applied; rather, it occurs at a point that is 90° later in the direction of rotation from where the pressure was applied. This principle allows the gyro to determine a rate of turn by sensing the amount of pressure created by a change in direction. Precession can also create some minor errors in some instruments.

ATTITUDE INDICATOR
The attitude indicator provides a substitute for the natural horizon. It is the only instrument that provides an immediate and direct indication of the helicopter's pitch and bank attitude. Since most attitude indicators

installed in helicopters are electrically powered, there may be a separate power switch, as well as a warning flag within the instrument, that indicates a loss of power. A caging or "quick erect" knob may be included, so you can stabilize the spin axis if the gyro has tumbled. [Figure 12-5]

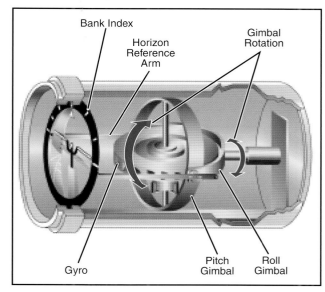

Figure 12-5. The gyro in the attitude indicator spins in the horizontal plane. Two mountings, or gimbals, are used so that both pitch and roll can be sensed simultaneously. Due to rigidity in space, the gyro remains in a fixed position relative to the horizon as the case and helicopter rotate around it.

HEADING INDICATOR
The heading indicator, which is sometimes referred to as a directional gyro (DG), senses movement around the vertical axis and provides a more accurate heading reference compared to a magnetic compass, which has a number of turning errors. [Figure 12-6].

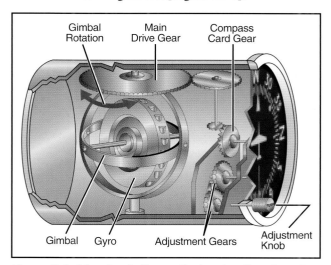

Figure 12-6. A heading indicator displays headings based on a 360° azimuth, with the final zero omitted. For example, a 6 represents 060°, while a 21 indicates 210°. The adjustment knob is used to align the heading indicator with the magnetic compass.

Due to internal friction within the gyroscope, precession is common in heading indicators. Precession causes the selected heading to drift from the set value. Some heading indicators receive a magnetic north reference from a remote source and generally need no adjustment. Heading indicators that do not have this automatic north-seeking capability are often called "free" gyros, and require that you periodically adjust them. You should align the heading indicator with the magnetic compass before flight and check it at 15-minute intervals during flight. When you do an in-flight alignment, be certain you are in straight-and-level, unaccelerated flight, with the magnetic compass showing a steady indication.

TURN INDICATORS

Turn indicators show the direction and the rate of turn. A standard rate turn is 3° per second, and at this rate you will complete a 360° turn in two minutes. A half-standard rate turn is 1.5° per second. Two types of indicators are used to display this information. The turn-and-slip indicator uses a needle to indicate direction and turn rate. When the needle is aligned with the white markings, called the turn index, you are in a standard rate turn. A half-standard rate turn is indicated when the needle is halfway between the indexes. The turn-and-slip indicator does not indicate roll rate. The turn coordinator is similar to the turn-and-slip indicator, but the gyro is canted, which allows it to sense roll rate in addition to rate of turn. The turn coordinator uses a miniature aircraft to indicate direction, as well as the turn and roll rate. [Figure 12-7]

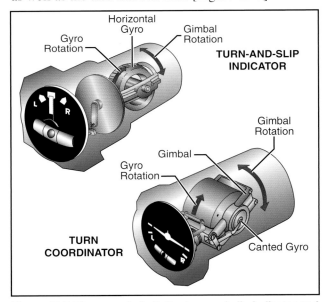

Figure 12-7. The gyros in both the turn-and-slip indicator and the turn coordinator are mounted so that they rotate in a vertical plane. The gimbal in the turn coordinator is set at an angle, or canted, which means precession allows the gyro to sense both rate of roll and rate of turn. The gimbal in the turn-and-slip indicator is horizontal. In this case, precession allows the gyro to sense only rate of turn. When the needle or miniature aircraft is aligned with the turn index, you are in a standard-rate turn.

Another part of both the turn coordinator and the turn-and-slip indicator is the inclinometer. The position of the ball defines whether the turn is coordinated or not. The helicopter is either slipping or skidding anytime the ball is not centered, and usually requires an adjustment of the antitorque pedals or angle of bank to correct it. [Figure 12-8]

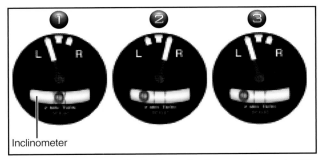

Figure 12-8. In a coordinated turn (instrument 1), the ball is centered. In a skid (instrument 2), the rate of turn is too great for the angle of bank, and the ball moves to the outside of the turn. Conversely, in a slip (instrument 3), the rate of turn is too small for the angle of bank, and the ball moves to the inside of the turn.

INSTRUMENT CHECK—During your preflight, check to see that the inclinometer is full of fluid and has no air bubbles. The ball should also be resting at its lowest point. Since almost all gyroscopic instruments installed in a helicopter are electrically driven, check to see that the power indicators are displaying off indications. Turn the master switch on and listen to the gyros spool up. There should be no abnormal sounds, such as a grinding sound, and the power out indicator flags should not be displayed. After engine start and before liftoff, set the direction indicator to the magnetic compass. During hover turns, check the heading indicator for proper operation and ensure that it has not precessed significantly. The turn indicator should also indicate a turn in the correct direction. During takeoff, check the attitude indicator for proper indication and recheck it during the first turn.

MAGNETIC COMPASS

In some helicopters, the magnetic compass is the only direction seeking instrument. Although the compass appears to move, it is actually mounted in such a way that the helicopter turns about the compass card as the card maintains its alignment with magnetic north.

COMPASS ERRORS

The magnetic compass can only give you reliable directional information if you understand its limitations and inherent errors. These include magnetic variation, compass deviation, and magnetic dip.

MAGNETIC VARIATION

When you fly under visual flight rules, you ordinarily navigate by referring to charts, which are oriented to true north. Because the aircraft compass is oriented to magnetic north, you must make allowances for the difference between these poles in order to navigate properly. You do this by applying a correction called variation to convert a true direction to a magnet direction. Variation at a given point is the angular difference between the true and magnetic poles. The amount of variation depends on where you are located on the earth's surface. Isogonic lines connect points where the variation is equal, while the agonic line defines the points where the variation is zero. [Figure 12-9]

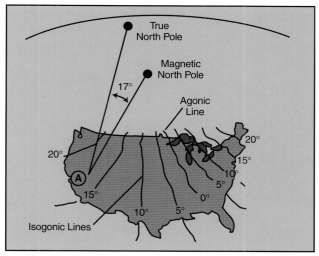

Figure 12-9. Variation at point A in the western United States is 17°. Since the magnetic north pole is located to the east of the true north pole in relation to this point, the variation is easterly. When the magnetic pole falls to the west of the true north pole, variation is westerly.

COMPASS DEVIATION
Besides the magnetic fields generated by the earth, other magnetic fields are produced by metal and electrical accessories within the helicopter. These magnetic fields distort the earth's magnet force and cause the compass to swing away from the correct heading. Manufacturers often install compensating magnets within the compass housing to reduce the effects of deviation. These magnets are usually adjusted while the engine is running and all electrical equipment is operating. Deviation error, however, cannot be completely eliminated; therefore, a compass correction card is mounted near the compass. The compass correction card corrects for deviation that occurs from one heading to the next as the lines of force interact at different angles.

MAGNETIC DIP
Magnetic dip is the result of the vertical component of the earth's magnetic field. This dip is virtually non-existent at the magnetic equator, since the lines of force are parallel to the earth's surface and the vertical component is minimal. As you move a compass toward the poles, the vertical component increases, and magnetic dip becomes more apparent at these higher latitudes.

Magnetic dip is responsible for compass errors during acceleration, deceleration, and turns.

Acceleration and deceleration errors are fluctuations in the compass during changes in speed. In the northern hemisphere, the compass swings toward the north during acceleration and toward the south during deceleration. When the speed stabilizes, the compass returns to an accurate indication. This error is most pronounced when you are flying on a heading of east or west, and decreases gradually as you fly closer to a north or south heading. The error does not occur when you are flying directly north or south. The memory aid, ANDS (Accelerate North, Decelerate South) may help you recall this error. In the southern hemisphere, this error occurs in the opposite direction.

Turning errors are most apparent when you are turning to or from a heading of north or south. This error increases as you near the poles as magnetic dip becomes more apparent. There is no turning error when flying near the magnetic equator. In the northern hemisphere, when you make a turn from a northerly heading, the compass gives an initial indication of a turn in the opposite direction. It then begins to show the turn in the proper direction, but lags behind the actual heading. The amount of lag decreases as the turn continues, then disappears as the helicopter reaches a heading of east or west. When you make a turn from a southerly heading, the compass gives an indication of a turn in the correct direction, but leads the actual heading. This error also disappears as the helicopter approaches an east or west heading.

INSTRUMENT CHECK—Prior to flight, make sure that the compass is full of fluid. During hover turns, the compass should swing freely and indicate known headings. Since that magnetic compass is required for all flight operations, the aircraft should never be flown with a faulty compass.

INSTRUMENT FLIGHT
To achieve smooth, positive control of the helicopter during instrument flight, you need to develop three fundamental skills. They are instrument cross-check, instrument interpretation, and aircraft control.

INSTRUMENT CROSS-CHECK
Cross-checking, sometimes referred to as scanning, is the continuous and logical observation of instruments for attitude and performance information. In attitude instrument flying, an attitude is maintained by reference to the instruments, which produces the desired result in performance. Due to human error, instrument error, and helicopter performance differences in various atmospheric and loading conditions, it is difficult to establish an attitude and have performance remain constant for a long period of time. These variables make

it necessary for you to constantly check the instruments and make appropriate changes in the helicopter's attitude. The actual technique may vary depending on what instruments are installed and where they are installed, as well as your experience and proficiency level. For this discussion, we will concentrate on the six basic flight instruments discussed earlier. [Figure 12-10]

At first, you may have a tendency to cross-check rapidly, looking directly at the instruments without knowing exactly what information you are seeking. However, with familiarity and practice, the instrument cross-check reveals definite trends during specific flight conditions. These trends help you control the helicopter as it makes a transition from one flight condition to another.

If you apply your full concentration to a single instrument, you will encounter a problem called "fixation." This results from a natural human inclination to observe a specific instrument carefully and accurately, often to the exclusion of other instruments. Fixation on a single instrument usually results in poor control. For example, while performing a turn, you may have a tendency to watch only the turn-and-slip indicator instead of including other instruments in your cross-check. This fixation on the turn-and-slip indicator often leads to a loss of altitude through poor pitch and bank control. You should look at each instrument only long enough to understand the information it presents, then continue on to the next one. Similarly, you may find yourself placing too much "emphasis" on a single instrument, instead of relying on a combination of instruments necessary for helicopter performance information. This differs from fixation in that you are using other instruments, but are giving too much attention to a particular one.

During performance of a maneuver, you may sometimes fail to anticipate significant instrument indications following attitude changes. For example, during leveloff from a climb or descent, you may concentrate on pitch control, while forgetting about heading or roll information. This error, called "omission," results in erratic control of heading and bank.

In spite of these common errors, most pilots can adapt well to flight by instrument reference after instruction and practice. You may find that you can control the helicopter more easily and precisely by instruments.

INSTRUMENT INTERPRETATION

The flight instruments together give a picture of what is going on. No one instrument is more important than the next; however, during certain maneuvers or conditions, those instruments that provide the most pertinent and useful information are termed primary instruments. Those which back up and supplement the primary instruments are termed supporting instruments. For example, since the attitude indicator is the only instrument that provides instant and direct aircraft attitude information, it should be considered primary during any change in pitch or bank attitude. After the new attitude is established, other instruments become primary, and the attitude indicator usually becomes the supporting instrument.

Figure 12-10. In most situations, the cross-check pattern includes the attitude indicator between the cross-check of each of the other instruments. A typical cross-check might progress as follows: attitude indicator, altimeter, attitude indicator, VSI, attitude indicator, heading indicator, attitude indicator, and so on.

AIRCRAFT CONTROL

Controlling the helicopter is the result of accurately interpreting the flight instruments and translating these readings into correct control responses. Aircraft control involves adjustment to pitch, bank, power, and trim in order to achieve a desired flight path.

Pitch attitude control is controlling the movement of the helicopter about its lateral axis. After interpreting the helicopter's pitch attitude by reference to the pitch instruments (attitude indicator, altimeter, airspeed indicator, and vertical speed indicator), cyclic control adjustments are made to affect the desired pitch attitude. In this chapter, the pitch attitudes illustrated are approximate and will vary with different helicopters.

Bank attitude control is controlling the angle made by the lateral tilt of the rotor and the natural horizon, or, the movement of the helicopter about its longitudinal axis. After interpreting the helicopter's bank instruments (attitude indicator, heading indicator, and turn indicator), cyclic control adjustments are made to attain the desired bank attitude.

Power control is the application of collective pitch with corresponding throttle control, where applicable. In straight-and-level flight, changes of collective pitch are made to correct for altitude deviations if the error is more than 100 feet, or the airspeed is off by more than 10 knots. If the error is less than that amount, use a slight cyclic climb or descent.

In order to fly a helicopter by reference to the instruments, you should know the approximate power settings required for your particular helicopter in various load configurations and flight conditions.

Trim, in helicopters, refers to the use of the cyclic centering button, if the helicopter is so equipped, to relieve all possible cyclic pressures. Trim also refers to the use of pedal adjustment to center the ball of the turn indicator. Pedal trim is required during all power changes.

The proper adjustment of collective pitch and cyclic friction helps you relax during instrument flight. Friction should be adjusted to minimize overcontrolling and to prevent creeping, but not applied to such a degree that control movement is limited. In addition, many helicopters equipped for instrument flight contain stability augmentation systems or an autopilot to help relieve pilot workload.

STRAIGHT-AND-LEVEL FLIGHT

Straight-and-level unaccelerated flight consists of maintaining the desired altitude, heading, airspeed, and pedal trim.

PITCH CONTROL

The pitch attitude of a helicopter is the angular relation of its longitudinal axis and the natural horizon. If available, the attitude indicator is used to establish the desired pitch attitude. In level flight, pitch attitude varies with airspeed and center of gravity. At a constant altitude and a stabilized airspeed, the pitch attitude is approximately level. [Figure 12-11]

Figure 12-11. The flight instruments for pitch control are the airspeed indicator, attitude indicator, altimeter, and vertical speed indicator.

ATTITUDE INDICATOR

The attitude indicator gives a direct indication of the pitch attitude of the helicopter. In visual flight, you attain the desired pitch attitude by using the cyclic to raise and lower the nose of the helicopter in relation to the natural horizon. During instrument flight, you follow exactly the same procedure in raising or lowering the miniature aircraft in relation to the horizon bar.

You may note some delay between control application and resultant instrument change. This is the normal control lag in the helicopter and should not be confused with instrument lag. The attitude indicator may show small misrepresentations of pitch attitude during maneuvers involving acceleration, deceleration, or turns. This precession error can be detected quickly by cross-checking the other pitch instruments.

If the miniature aircraft is properly adjusted on the ground, it may not require readjustment in flight. If the miniature aircraft is not on the horizon bar after level-off at normal cruising airspeed, adjust it as necessary while maintaining level flight with the other pitch instruments. Once the miniature aircraft has been adjusted in level flight at normal cruising airspeed, leave it unchanged so it will give an accurate picture of pitch attitude at all times.

When making initial pitch attitude corrections to maintain altitude, the changes of attitude should be small and smoothly applied. The initial movement of the horizon bar should not exceed one bar width high or low. [Figure 12-12] If a further change is required, an additional correction of one-half bar normally corrects any deviation from the desired altitude. This one and

Figure 12-12. The initial pitch correction at normal cruise is one bar width.

one-half bar correction is normally the maximum pitch attitude correction from level flight attitude. After you have made the correction, cross-check the other pitch instruments to determine whether the pitch attitude change is sufficient. If more correction is needed to return to altitude, or if the airspeed varies more than 10 knots from that desired, adjust the power.

ALTIMETER

The altimeter gives an indirect indication of the pitch attitude of the helicopter in straight-and-level flight. Since the altitude should remain constant in level flight, deviation from the desired altitude shows a need for a change in pitch attitude, and if necessary, power. When losing altitude, raise the pitch attitude and, if necessary, add power. When gaining altitude, lower the pitch attitude and, if necessary, reduce power.

The rate at which the altimeter moves helps in determining pitch attitude. A very slow movement of the altimeter indicates a small deviation from the desired pitch attitude, while a fast movement of the altimeter indicates a large deviation from the desired pitch attitude. Make any corrective action promptly, with small control changes. Also, remember that movement of the altimeter should always be corrected by two distinct changes. The first is a change of attitude to stop the altimeter; and the second, a change of attitude to return smoothly to the desired altitude. If the altitude and airspeed are more than 100 feet and 10 knots low, respectively, apply power along with an increase of pitch attitude. If the altitude and airspeed are high by more than 100 feet and 10 knots, reduce power and lower the pitch attitude.

There is a small lag in the movement of the altimeter; however, for all practical purposes, consider that the altimeter gives an immediate indication of a change, or a need for change in pitch attitude.

Since the altimeter provides the most pertinent information regarding pitch in level flight, it is considered primary for pitch.

VERTICAL SPEED INDICATOR

The vertical speed indicator gives an indirect indication of the pitch attitude of the helicopter and should be used in conjunction with the other pitch instruments to attain a high degree of accuracy and precision. The instrument indicates zero when in level flight. Any movement of the needle from the zero position shows a need for an immediate change in pitch attitude to return it to zero. Always use the vertical speed indicator in conjunction with the altimeter in level flight. If a movement of the vertical speed indicator is detected, immediately use the proper corrective measures to return it to zero. If the correction is made promptly, there is usually little or no change in altitude. If you do not zero the needle of the vertical speed indicator immediately, the results will show on the altimeter as a gain or loss of altitude.

The initial movement of the vertical speed needle is instantaneous and indicates the trend of the vertical movement of the helicopter. It must be realized that a period of time is necessary for the vertical speed indicator to reach its maximum point of deflection after a correction has been made. This time element is commonly referred to as "lag." The lag is directly proportional to the speed and magnitude of the pitch change. If you employ smooth control techniques and make small adjustments in pitch attitude, lag is minimized, and the vertical speed indicator is easy to interpret. Overcontrolling can be minimized by first neutralizing the controls and allowing the pitch attitude to stabilize; then readjusting the pitch attitude by noting the indications of the other pitch instruments.

Occasionally, the vertical speed indicator may be slightly out of calibration. This could result in the instrument indicating a slight climb or descent even when the helicopter is in level flight. If it cannot be readjusted properly, this error must be taken into consideration when using the vertical speed indicator for pitch control. For example, if the vertical speed indicator showed a descent of 100 f.p.m. when the helicopter was in level flight, you would have to use that indication as level flight. Any deviation from that reading would indicate a change in attitude.

AIRSPEED INDICATOR

The airspeed indicator gives an indirect indication of helicopter pitch attitude. With a given power setting and pitch attitude, the airspeed remains constant. If the airspeed increases, the nose is too low and should be raised. If the airspeed decreases, the nose is too high and should be lowered. A rapid change in airspeed indicates a large change in pitch attitude, and a slow change in airspeed indicates a small change in pitch attitude. There is very little lag in the indications of the airspeed indicator. If, while making attitude changes, you notice some lag between control application and change of airspeed, it is most likely due to cyclic control lag. Generally, a departure from the desired airspeed, due to an inadvertent pitch attitude change, also results in a change in altitude. For example, an increase in airspeed due to a low pitch attitude results in a decrease in altitude. A correction in the pitch attitude regains both airspeed and altitude.

BANK CONTROL

The bank attitude of a helicopter is the angular relation of its lateral axis and the natural horizon. To maintain a straight course in visual flight, you must keep the lateral axis of the helicopter level with the natural horizon. Assuming the helicopter is in coordinated flight, any deviation from a laterally level attitude produces a turn. [Figure 12-13]

ATTITUDE INDICATOR

The attitude indicator gives a direct indication of the bank attitude of the helicopter. For instrument flight,

Figure 12-13. The flight instruments used for bank control are the attitude, heading, and turn indicators.

the miniature aircraft and the horizon bar of the attitude indicator are substituted for the actual helicopter and the natural horizon. Any change in bank attitude of the helicopter is indicated instantly by the miniature aircraft. For proper interpretations of this instrument, you should imagine being in the miniature aircraft. If the helicopter is properly trimmed and the rotor tilts, a turn begins. The turn can be stopped by leveling the miniature aircraft with the horizon bar. The ball in the turn-and-slip indicator should always be kept centered through proper pedal trim.

The angle of bank is indicated by the pointer on the banking scale at the top of the instrument. [Figure 12-14] Small bank angles, which may not be seen by observing the miniature aircraft, can easily be determined by referring to the banking scale pointer.

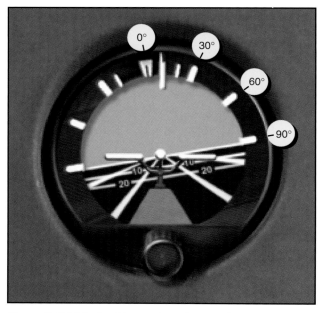

Figure 12-14. The banking scale at the top of the attitude indicator indicates varying degrees of bank. In this example, the helicopter is banked a little over 10° to the right.

Pitch and bank attitudes can be determined simultaneously on the attitude indicator. Even though the miniature aircraft is not level with the horizon bar, pitch attitude can be established by observing the relative position of the miniature aircraft and the horizon bar.

The attitude indicator may show small misrepresentations of bank attitude during maneuvers that involve turns. This precession error can be immediately detected by closely cross-checking the other bank instruments during these maneuvers. Precession normally is noticed when rolling out of a turn. If, on the completion of a turn, the miniature aircraft is level and

the helicopter is still turning, make a small change of bank attitude to center the turn needle and stop the movement of the heading indicator.

HEADING INDICATOR

In coordinated flight, the heading indicator gives an indirect indication of the helicopter's bank attitude. When a helicopter is banked, it turns. When the lateral axis of the helicopter is level, it flies straight. Therefore, in coordinated flight, when the heading indicator shows a constant heading, the helicopter is level laterally. A deviation from the desired heading indicates a bank in the direction the helicopter is turning. A small angle of bank is indicated by a slow change of heading; a large angle of bank is indicated by a rapid change of heading. If a turn is noticed, apply opposite cyclic until the heading indicator indicates the desired heading, simultaneously checking that the ball is centered. When making the correction to the desired heading, you should not use a bank angle greater than that required to achieve a standard rate turn. In addition, if the number of degrees of change is small, limit the bank angle to the number of degrees to be turned. Bank angles greater than these require more skill and precision in attaining the desired results. During straight-and-level flight, the heading indicator is the primary reference for bank control.

TURN INDICATOR

During coordinated flight, the needle of the turn-and-slip indicator gives an indirect indication of the bank attitude of the helicopter. When the needle is displaced from the vertical position, the helicopter is turning in the direction of the displacement. Thus, if the needle is displaced to the left, the helicopter is turning left. Bringing the needle back to the vertical position with the cyclic produces straight flight. A close observation of the needle is necessary to accurately interpret small deviations from the desired position.

Cross-check the ball of the turn-and-slip indicator to determine that the helicopter is in coordinated flight. If the rotor is laterally level and torque is properly compensated for by pedal pressure, the ball remains in the center. To center the ball, level the helicopter laterally by reference to the other bank instruments, then center the ball with pedal trim. Torque correction pressures vary as you make power changes. Always check the ball following such changes.

COMMON ERRORS DURING STRAIGHT-AND-LEVEL FLIGHT

1. Failure to maintain altitude.
2. Failure to maintain heading.
3. Overcontrolling pitch and bank during corrections.
4. Failure to maintain proper pedal trim.
5. Failure to cross-check all available instruments.

POWER CONTROL DURING STRAIGHT-AND-LEVEL FLIGHT

Establishing specific power settings is accomplished through collective pitch adjustments and throttle control, where necessary. For reciprocating powered helicopters, power indications are observed on the manifold pressure gauge. For turbine powered helicopters, power is observed on the torque gauge. (Since most IFR certified helicopters are turbine powered, this discussion concentrates on this type of helicopter.)

At any given airspeed, a specific power setting determines whether the helicopter is in level flight, in a climb, or in a descent. For example, cruising airspeed maintained with cruising power results in level flight. If you increase the power setting and hold the airspeed constant, the helicopter climbs. Conversely, if you decrease power and hold the airspeed constant, the helicopter descends. As a rule of thumb, in a turbine-engine powered helicopter, a 10 to 15 percent change in the torque value required to maintain level flight results in a climb or descent of approximately 500 f.p.m., if the airspeed remains the same.

If the altitude is held constant, power determines the airspeed. For example, at a constant altitude, cruising power results in cruising airspeed. Any deviation from the cruising power setting results in a change of airspeed. When power is added to increase airspeed, the nose of the helicopter pitches up and yaws to the right in a helicopter with a counterclockwise main rotor blade rotation. When power is reduced to decrease airspeed, the nose pitches down and yaws to the left. The yawing effect is most pronounced in single-rotor helicopters, and is absent in helicopters with counter-rotating rotors. To counteract the yawing tendency of the helicopter, apply pedal trim during power changes.

To maintain a constant altitude and airspeed in level flight, coordinate pitch attitude and power control. The relationship between altitude and airspeed determines the need for a change in power and/or pitch attitude. If the altitude is constant and the airspeed is high or low, change the power to obtain the desired airspeed. During the change in power, make an accurate interpretation of the altimeter; then counteract any deviation from the desired altitude by an appropriate change of pitch attitude. If the altitude is low and the airspeed is high, or vice versa, a change in pitch attitude alone may return the helicopter to the proper altitude and airspeed. If both airspeed and altitude are low, or if both are high, a change in both power and pitch attitude is necessary.

To make power control easy when changing airspeed, it is necessary to know the approximate power settings for the various airspeeds that will be flown. When the airspeed is to be changed any appreciable amount, adjust the torque so that it is approximately five percent over or under that setting necessary to maintain the new airspeed. As the power approaches the desired setting, include the torque meter in the cross-check to determine when the proper adjustment has been accomplished. As the airspeed is changing, adjust the pitch attitude to maintain a constant altitude. A constant heading should be maintained throughout the change. As the desired airspeed is approached, adjust power to the new cruising power setting and further adjust pitch attitude to maintain altitude. Overpowering and underpowering torque approximately five percent results in a change of airspeed at a moderate rate, which allows ample time to adjust pitch and bank smoothly. The instrument indications for straight-and-level flight at normal cruise, and during the transition from normal cruise to slow cruise are illustrated in figures 12-15 and 12-16 on the next page. After the airspeed has stabilized at slow cruise, the attitude indicator shows an approximate level pitch attitude.

The altimeter is the primary pitch instrument during level flight, whether flying at a constant airspeed, or during a change in airspeed. Altitude should not change during airspeed transitions. The heading indicator remains the primary bank instrument. Whenever the airspeed is changed any appreciable amount, the torque meter is momentarily the primary instrument for power control. When the airspeed approaches that desired, the airspeed indicator again becomes the primary instrument for power control.

The cross-check of the pitch and bank instruments to produce straight-and-level flight should be combined with the power control instruments. With a constant power setting, a normal cross-check should be satisfactory. When changing power, the speed of the cross-check must be increased to cover the pitch and bank instruments adequately. This is necessary to counteract any deviations immediately.

COMMON ERRORS DURING AIRSPEED CHANGES
1. Improper use of power.
2. Overcontrolling pitch attitude.
3. Failure to maintain heading.
4. Failure to maintain altitude.
5. Improper pedal trim.

STRAIGHT CLIMBS (CONSTANT AIRSPEED AND CONSTANT RATE)

For any power setting and load condition, there is only one airspeed that will give the most efficient rate of climb. To determine this, you should consult the climb data for the type of helicopter being flown. The technique varies according to the airspeed on entry and whether you want to make a constant airspeed or constant rate climb.

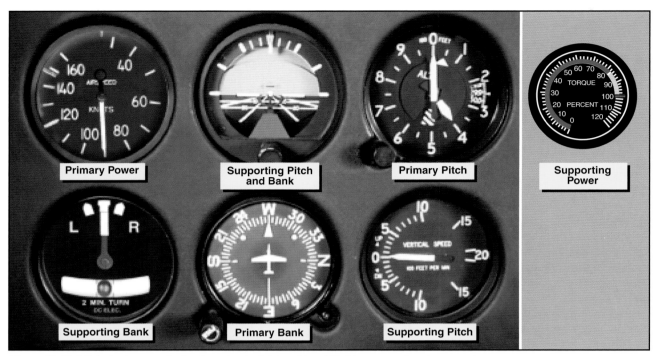

Figure 12-15. Flight instrument indications in straight-and-level flight at normal cruise speed.

ENTRY

To enter a constant airspeed climb from cruise airspeed, when the climb speed is lower than cruise speed, simultaneously increase power to the climb power setting and adjust pitch attitude to the approximate climb attitude. The increase in power causes the helicopter to start climbing and only very slight back cyclic pressure is needed to complete the change from level to climb attitude. The attitude indicator should be used to accomplish the pitch change. If the transition from level flight to a climb is smooth, the vertical speed indicator shows an immediate upward trend and then stops at a rate appropriate to the stabilized airspeed and attitude. Primary and supporting instruments for climb entry are illustrated in figure 12-17.

When the helicopter stabilizes on a constant airspeed and attitude, the airspeed indicator becomes primary

Figure 12-16. Flight instrument indications in straight-and-level flight with airspeed decreasing.

Figure 12-17. Flight instrument indications during climb entry for a constant airspeed climb.

for pitch. The torque meter continues to be primary for power and should be monitored closely to determine if the proper climb power setting is being maintained. Primary and supporting instruments for a stabilized constant airspeed climb are shown in figure 12-18.

The technique and procedures for entering a constant rate climb are very similar to those previously described for a constant airspeed climb. For training purposes, a constant rate climb is entered from climb airspeed. The rate used is the one that is appropriate for

Figure 12-18. Flight instrument indications in a stabilized, constant airspeed climb.

the particular helicopter being flown. Normally, in helicopters with low climb rates, 500 f.p.m. is appropriate, in helicopters capable of high climb rates, use a rate of 1,000 f.p.m.

To enter a constant rate climb, increase power to the approximate setting for the desired rate. As power is applied, the airspeed indicator is primary for pitch until the vertical speed approaches the desired rate. At this time, the vertical speed indicator becomes primary for pitch. Change pitch attitude by reference to the attitude indicator to maintain the desired vertical speed. When the VSI becomes primary for pitch, the airspeed indicator becomes primary for power. Primary and supporting instruments for a stabilized constant rate climb are illustrated in figure 12-19. Adjust power to maintain desired airspeed. Pitch attitude and power corrections should be closely coordinated. To illustrate this, if the vertical speed is correct but the airspeed is low, add power. As power is increased, it may be necessary to lower the pitch attitude slightly to avoid increasing the vertical rate. Adjust the pitch attitude smoothly to avoid overcontrolling. Small power corrections usually will be sufficient to bring the airspeed back to the desired indication.

LEVELOFF

The leveloff from a constant airspeed climb must be started before reaching the desired altitude. Although the amount of lead varies with the helicopter being flown and your piloting technique, the most important factor is vertical speed. As a rule of thumb, use 10 percent of the vertical velocity as your lead point. For example, if the

rate of climb is 500 f.p.m., initiate the leveloff approximately 50 feet before the desired altitude. When the proper lead altitude is reached, the altimeter becomes primary for pitch. Adjust the pitch attitude to the level flight attitude for that airspeed. Cross-check the altimeter and VSI to determine when level flight has been attained at the desired altitude. To level off at cruise airspeed, if this speed is higher than climb airspeed, leave the power at the climb power setting until the airspeed approaches cruise airspeed, then reduce it to the cruise power setting.

The leveloff from a constant rate climb is accomplished in the same manner as the leveloff from a constant airspeed climb.

STRAIGHT DESCENTS (CONSTANT AIRSPEED AND CONSTANT RATE)

A descent may be performed at any normal airspeed the helicopter is capable of, but the airspeed must be determined prior to entry. The technique is determined by whether you want to perform a constant airspeed or a constant rate descent.

ENTRY

If your airspeed is higher than descending airspeed, and you wish to make a constant airspeed descent at the descending airspeed, reduce power to the descending power setting and maintain a constant altitude using cyclic pitch control. When you approach the descending airspeed, the airspeed indicator becomes primary for pitch, and the torque meter is primary for power. As you hold the airspeed constant, the helicopter begins to descend. For a constant rate descent, reduce the power

Figure 12-19. Flight instrument indications in a stabilized constant rate climb.

to the approximate setting for the desired rate. If the descent is started at the descending airspeed, the airspeed indicator is primary for pitch until the VSI approaches the desired rate. At this time, the vertical speed indicator becomes primary for pitch, and the airspeed indicator becomes primary for power. Coordinate power and pitch attitude control as was described earlier for constant rate climbs.

LEVELOFF

The leveloff from a constant airspeed descent may be made at descending airspeed or at cruise airspeed, if this is higher than descending airspeed. As in a climb leveloff, the amount of lead depends on the rate of descent and control technique. For a leveloff at descending airspeed, the lead should be approximately 10 percent of the vertical speed. At the lead altitude, simultaneously increase power to the setting necessary to maintain descending airspeed in level flight. At this point, the altimeter becomes primary for pitch, and the airspeed indicator becomes primary for power.

To level off at a higher airspeed than descending airspeed, increase the power approximately 100 to 150 feet prior to reaching the desired altitude. The power setting should be that which is necessary to maintain the desired airspeed in level flight. Hold the vertical speed constant until approximately 50 feet above the desired altitude. At this point, the altimeter becomes primary for pitch, and the airspeed indicator becomes primary for power. The leveloff from a constant rate descent should be accomplished in the same manner as the leveloff from a constant airspeed descent.

COMMON ERRORS DURING STRAIGHT CLIMBS AND DESCENTS

1. Failure to maintain heading.
2. Improper use of power.
3. Poor control of pitch attitude.
4. Failure to maintain proper pedal trim.
5. Failure to level off on desired altitude.

TURNS

When making turns by reference to the flight instruments, they should be made at a definite rate. Turns described in this chapter are those that do not exceed a standard rate of 3° per second as indicated on the turn-and-slip indicator. True airspeed determines the angle of bank necessary to maintain a standard rate turn. A rule of thumb to determine the approximate angle of bank required for a standard rate turn is to divide your airspeed by 10 and add one-half the result. For example, at 60 knots, approximately 9° of bank is required (60 ÷ 10 = 6 + 3 = 9); at 80 knots, approximately 12° of bank is needed for a standard rate turn.

To enter a turn, apply lateral cyclic in the direction of the desired turn. The entry should be accomplished smoothly, using the attitude indicator to establish the approximate bank angle. When the turn indicator indicates a standard rate turn, it becomes primary for bank. The attitude indicator now becomes a supporting instrument. During level turns, the altimeter is primary for pitch, and the airspeed indicator is primary for power. Primary and supporting instruments for a stabilized standard rate turn are illustrated in figure 12-20. If an

Figure 12-20. Flight instrument indications for a standard rate turn to the left.

increase in power is required to maintain airspeed, slight forward cyclic pressure may be required since the helicopter tends to pitch up as collective pitch angle is increased. Apply pedal trim, as required, to keep the ball centered.

To recover to straight-and-level flight, apply cyclic in the direction opposite the turn. The rate of roll-out should be the same as the rate used when rolling into the turn. As you initiate the turn recover, the attitude indicator becomes primary for bank. When the helicopter is approximately level, the heading indicator becomes primary for bank as in straight-and-level flight. Cross-check the airspeed indicator and ball closely to maintain the desired airspeed and pedal trim.

TURNS TO A PREDETERMINED HEADING
A helicopter turns as long as its lateral axis is tilted; therefore, the recovery must start before the desired heading is reached. The amount of lead varies with the rate of turn and your piloting technique.

As a guide, when making a 3° per second rate of turn, use a lead of one-half the bank angle. For example, if you are using a 12° bank angle, use half of that, or 6°, as the lead point prior to your desired heading. Use this lead until you are able to determine the exact amount required by your particular technique. The bank angle should never exceed the number of degrees to be turned. As in any standard rate turn, the rate of recovery should be the same as the rate for entry. During turns to predetermined headings, cross-check the primary and supporting pitch, bank, and power instruments closely.

TIMED TURNS
A timed turn is a turn in which the clock and turn-and-slip indicator are used to change heading a definite number of degrees in a given time. For example, using a standard rate turn, a helicopter turns 45° in 15 seconds. Using a half-standard rate turn, the helicopter turns 45° in 30 seconds. Timed turns can be used if your heading indicator becomes inoperative.

Prior to performing timed turns, the turn coordinator should be calibrated to determine the accuracy of its indications. To do this, establish a standard rate turn by referring to the turn-and-slip indicator. Then as the sweep second hand of the clock passes a cardinal point (12, 3, 6, or 9), check the heading on the heading indicator. While holding the indicated rate of turn constant, note the heading changes at 10-second intervals. If the helicopter turns more or less than 30° in that interval, a smaller or larger deflection of the needle is necessary to produce a standard rate turn. When you have calibrated the turn-and-slip indicator during turns in each direction, note the corrected deflections, if any, and apply them during all timed turns.

You use the same cross-check and control technique in making timed turns that you use to make turns to a predetermined heading, except that you substitute the clock for the heading indicator. The needle of the turn-and-slip indicator is primary for bank control, the altimeter is primary for pitch control, and the airspeed indicator is primary for power control. Begin the roll-in when the clock's second hand passes a cardinal point, hold the turn at the calibrated standard-rate indication, or half-standard-rate for small changes in heading, and begin the roll-out when the computed number of seconds has elapsed. If the roll-in and roll-out rates are the same, the time taken during entry and recovery need not be considered in the time computation.

If you practice timed turns with a full instrument panel, check the heading indicator for the accuracy of your turns. If you execute the turns without the heading indicator, use the magnetic compass at the completion of the turn to check turn accuracy, taking compass deviation errors into consideration.

CHANGE OF AIRSPEED IN TURNS
Changing airspeed in turns is an effective maneuver for increasing your proficiency in all three basic instrument skills. Since the maneuver involves simultaneous changes in all components of control, proper execution requires a rapid cross-check and interpretation, as well as smooth control. Proficiency in the maneuver also contributes to your confidence in the instruments during attitude and power changes involved in more complex maneuvers.

Pitch and power control techniques are the same as those used during airspeed changes in straight-and-level flight. As discussed previously, the angle of bank necessary for a given rate of turn is proportional to the true airspeed. Since the turns are executed at standard rate, the angle of bank must be varied in direct proportion to the airspeed change in order to maintain a constant rate of turn. During a reduction of airspeed, you must decrease the angle of bank and increase the pitch attitude to maintain altitude and a standard rate turn.

The altimeter and the needle on the turn indicator should remain constant throughout the turn. The altimeter is primary for pitch control, and the turn needle is primary for bank control. The torque meter is primary for power control while the airspeed is changing. As the airspeed approaches the new indication, the airspeed indicator becomes primary for power control.

Two methods of changing airspeed in turns may be used. In the first method, airspeed is changed after the turn is established. In the second method, the airspeed change is initiated simultaneously with the turn entry. The first method is easier, but regardless of the method used, the rate of cross-check must be increased as you reduce power. As the helicopter decelerates, check the altimeter and VSI for needed pitch changes, and the bank instruments for needed bank changes. If the needle of the turn-and-slip indicator shows a deviation from the desired deflection, change the bank. Adjust pitch attitude to maintain altitude. When the airspeed approaches that desired, the airspeed indicator becomes primary for power control. Adjust the torque meter to maintain the desired airspeed. Use pedal trim to ensure the maneuver is coordinated.

Until your control technique is very smooth, frequently cross-check the attitude indicator to keep from over-controlling and to provide approximate bank angles appropriate for the changing airspeeds.

30° BANK TURN

A turn using 30° of bank is seldom necessary, or advisable, in IMC, but it is an excellent maneuver to increase your ability to react quickly and smoothly to rapid changes of attitude. Even though the entry and recovery technique are the same as for any other turn, you will probably find it more difficult to control pitch because of the decrease in vertical lift as the bank increases. Also, because of the decrease in vertical lift, there is a tendency to lose altitude and/or airspeed. Therefore, to maintain a constant altitude and airspeed, additional power is required. You should not initiate a correction, however, until the instruments indicate the need for a correction. During the maneuver, note the need for a correction on the altimeter and vertical speed indicator, then check the indications on the attitude

indicator, and make the necessary adjustments. After you have made this change, again check the altimeter and vertical speed indicator to determine whether or not the correction was adequate.

CLIMBING AND DESCENDING TURNS

For climbing and descending turns, the techniques described earlier for straight climbs and descents and those for standard rate turns are combined. For practice, start the climb or descent and turn simultaneously. The primary and supporting instruments for a stabilized constant airspeed left climbing turn are illustrated in figure 12-21. The leveloff from a climbing or descending turn is the same as the leveloff from a straight climb or descent. To recover to straight-and-level flight, you may stop the turn and then level off, level off and then stop the turn, or simultaneously level off and stop the turn. During climbing and descending turns, keep the ball of the turn indicator centered with pedal trim.

COMPASS TURNS

The use of gyroscopic heading indicators make heading control very easy. However, if the heading indicator fails or your helicopter does not have one installed, you must use the magnetic compass for heading reference. When making compass-only turns, you need to adjust for the lead or lag created by acceleration and deceleration errors so that you roll out on the desired heading. When turning to a heading of north, the lead for the roll-out must include the number of degrees of your latitude plus the lead you normally use in recovery from turns. During a turn to a south heading, maintain the turn until the compass passes south the number

Figure 12-21. Flight instrument indications for a stabilized left climbing turn at a constant airspeed.

of degrees of your latitude, minus your normal roll-out lead. For example, when turning from an easterly direction to north, where the latitude is 30°, start the roll-out when the compass reads 037° (30° plus one-half the 15° angle of bank, or whatever amount is appropriate for your rate of roll-out). When turning from an easterly direction to south, start the roll-out when the magnetic compass reads 203° (180° plus 30° minus one-half the angle of bank). When making similar turns from a westerly direction, the appropriate points at which to begin your roll-out would be 323° for a turn to north, and 157° for a turn to south.

COMMON ERRORS DURING TURNS

1. Failure to maintain desired turn rate.
2. Failure to maintain altitude in level turns.
3. Failure to maintain desired airspeed.
4. Variation in the rate of entry and recovery.
5. Failure to use proper lead in turns to a heading.
6. Failure to properly compute time during timed turns.
7. Failure to use proper leads and lags during the compass turns.
8. Improper use of power.
9. Failure to use proper pedal trim.

UNUSUAL ATTITUDES

Any maneuver not required for normal helicopter instrument flight is an unusual attitude and may be caused by any one or a combination of factors, such as turbulence, disorientation, instrument failure, confusion, preoccupation with cockpit duties, carelessness in cross-checking, errors in instrument interpretation, or lack of proficiency in aircraft control. Due to the instability characteristics of the helicopter, unusual attitudes can be extremely critical. As soon as you detect an unusual attitude, make a recovery to straight-and-level flight as soon as possible with a minimum loss of altitude.

To recover from an unusual attitude, correct bank and pitch attitude, and adjust power as necessary. All components are changed almost simultaneously, with little lead of one over the other. You must be able to perform this task with and without the attitude indicator. If the helicopter is in a climbing or descending turn, correct bank, pitch, and power. The bank attitude should be corrected by referring to the turn-and-slip indicator and attitude indicator. Pitch attitude should be corrected by reference to the altimeter, airspeed indicator, VSI, and attitude indicator. Adjust power by referring to the airspeed indicator and torque meter.

Since the displacement of the controls used in recoveries from unusual attitudes may be greater than those for normal flight, take care in making adjustments as straight-and-level flight is approached. Cross-check the other instruments closely to avoid overcontrolling.

COMMON ERRORS DURING UNUSUAL ATTITUDE RECOVERIES

1. Failure to make proper pitch correction.
2. Failure to make proper bank correction.
3. Failure to make proper power correction.
4. Overcontrol of pitch and/or bank attitude.
5. Overcontrol of power.
6. Excessive loss of altitude.

EMERGENCIES

Emergencies under instrument flight are handled similarly to those occurring during VFR flight. A thorough knowledge of the helicopter and its systems, as well as good aeronautical knowledge and judgment, prepares you to better handle emergency situations. Safe operations begin with preflight planning and a thorough preflight. Plan your route of flight so that there are adequate landing sites in the event you have to make an emergency landing. Make sure you have all your resources, such as maps, publications, flashlights, and fire extinguishers readily available for use in an emergency.

During any emergency, you should first fly the aircraft. This means that you should make sure the helicopter is under control, including the determination of emergency landing sites. Then perform the emergency checklist memory items, followed by written items in the RFM. Once all these items are under control, you should notify ATC. Declare any emergency on the last assigned ATC frequency, or if one was not issued, transmit on the emergency frequency 121.5. Set the transponder to the emergency squawk code 7700. This code triggers an alarm or a special indicator in radar facilities.

Most in-flight emergencies, including low fuel and a complete electrical failure, require you to **land as soon as possible**. In the event of an electrical fire, turn all non-essential equipment off and **land immediately**. Some essential electrical instruments, such as the attitude indicator, may be required for a safe landing. A navigation radio failure may not require an immediate landing as long as the flight can continue safely. In this case, you should **land as soon as practical**. ATC may be able to provide vectors to a safe landing area. For the specific details on what to do during an emergency, you should refer to the RFM for the helicopter you are flying.

Land as soon as possible—Land without delay at the nearest suitable area, such as an open field, at which a safe approach and landing is assured.

Land immediately—The urgency of the landing is paramount. The primary consideration is to assure the survival of the occupants. Landing in trees, water, or other unsafe areas should be considered only as a last resort.

Land as soon as practical—The landing site and duration of flight are at the discretion of the pilot. Extended flight beyond the nearest approved landing area is not recommended.

AUTOROTATIONS

Both straight-ahead and turning autorotations should be practiced by reference to instruments. This training will ensure that you can take prompt corrective action to maintain positive aircraft control in the event of an engine failure.

To enter autorotation, reduce collective pitch smoothly to maintain a safe rotor r.p.m. and apply pedal trim to keep the ball of the turn-and-slip indicator centered. The pitch attitude of the helicopter should be approximately level as shown by the attitude indicator. The airspeed indicator is the primary pitch instrument and should be adjusted to the recommended autorotation speed. The heading indicator is primary for bank in a straight-ahead autorotation. In a turning autorotation, a standard rate turn should be maintained by reference to the needle of the turn-and-slip indicator.

COMMON ERRORS DURING AUTOROTATIONS

1. Uncoordinated entry due to improper pedal trim.
2. Poor airspeed control due to improper pitch attitude.
3. Poor heading control in straight-ahead autorotations.
4. Failure to maintain proper rotor r.p.m.
5. Failure to maintain a standard rate turn during turning autorotations.

SERVO FAILURE

Most helicopters certified for single-pilot IFR flight are required to have autopilots, which greatly reduces pilot workload. If an autopilot servo fails, however, you have to resume manual control of the helicopter. How much your workload increases, depends on which servo fails. If a cyclic servo fails, you may want to land immediately as the workload increases tremendously.

If an antitorque or collective servo fails, you might be able to continue to the next suitable landing site.

INSTRUMENT TAKEOFF

This maneuver should only be performed as part of your training for an instrument rating. The procedures and techniques described here should be modified, as necessary, to conform with those set forth in the operating instructions for the particular helicopter being flown.

Adjust the miniature aircraft in the attitude indicator, as appropriate, for the aircraft being flown. After the helicopter is aligned with the runway or takeoff pad, to prevent forward movement of a helicopter equipped with a wheel-type landing gear, set the parking brake or apply the toe brakes. If the parking brake is used, it must be unlocked after the takeoff has been completed. Apply sufficient friction to the collective pitch control to minimize overcontrolling and to prevent creeping. Excessive friction should be avoided since this limits collective pitch movement.

After checking all instruments for proper indications, start the takeoff by applying collective pitch and a predetermined power setting. Add power smoothly and steadily to gain airspeed and altitude simultaneously and to prevent settling to the ground. As power is applied and the helicopter becomes airborne, use the antitorque pedals initially to maintain the desired heading. At the same time, apply forward cyclic to begin accelerating to climbing airspeed. During the initial acceleration, the pitch attitude of the helicopter, as read on the attitude indicator, should be one to two bar widths low. The primary and supporting instruments after becoming airborne are illustrated in figure 12-22. As the airspeed increases

Figure 12-22. Flight instrument indications during an instrument takeoff.

to the appropriate climb airspeed, adjust pitch gradually to climb attitude. As climb airspeed is reached, reduce power to the climb power setting and transition to a fully coordinated straight climb.

During the initial climbout, minor heading corrections should be made with pedals only until sufficient airspeed is attained to transition to fully coordinated flight. Throughout the instrument take-off, instrument cross-check and interpretations must be rapid and accurate, and aircraft control positive and smooth.

COMMON ERRORS DURING INSTRUMENT TAKEOFFS

1. Failure to maintain heading.
2. Overcontrolling pedals.
3. Failure to use required power.
4. Failure to adjust pitch attitude as climbing airspeed is reached.

Flying at night can be a very pleasant experience. The air is generally cooler and smoother, resulting in better helicopter performance and a more comfortable flight. You generally also experience less traffic and less radio congestion.

NIGHT FLIGHT PHYSIOLOGY

Before discussing night operations, it is important you understand how your vision is affected at night and how to counteract the visual illusions, which you might encounter.

VISION IN FLIGHT

Vision is by far the most important sense that you have, and flying is obviously impossible without it. Most of the things you perceive while flying are visual or heavily supplemented by vision. The visual sense is especially important in collision avoidance and depth perception. Your vision sensors are your eyes, even though they are not perfect in the way they function or see objects. Since your eyes are not always able to see all things at all times, illusions and blindspots occur. The more you understand the eye and how it functions, the easier it is to compensate for these illusions and blindspots.

THE EYE

The eye works in much the same way as a camera. Both have an aperture, lens, method of focusing, and a surface for registering images. [Figure 13-1].

Vision is primarily the result of light striking a photosensitive layer, called the retina, at the back of the eye. The retina is composed of light-sensitive cones and rods. The cones in your eye perceive an image best when the light is bright, while the rods work best in low light. The pattern of light that strikes the cones and rods is transmitted as electrical impulses by the optic nerve to the brain where these signals are interpreted as an image. The area where the optic nerve meets the retina contains no cones or rods, creating a blind spot in vision. Normally, each eye compensates for the other's blind spot. [Figure 13-2]

CONES

Cones are concentrated around the center of the retina. They gradually diminish in number as the distance from the center increases. Cones allow you to perceive color by sensing red, blue, and green light.

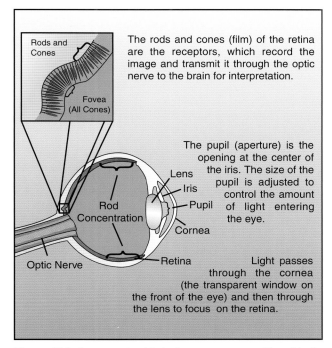

The rods and cones (film) of the retina are the receptors, which record the image and transmit it through the optic nerve to the brain for interpretation.

The pupil (aperture) is the opening at the center of the iris. The size of the pupil is adjusted to control the amount of light entering the eye.

Light passes through the cornea (the transparent window on the front of the eye) and then through the lens to focus on the retina.

Figure 13-1. A camera is able to focus on near and far objects by changing the distance between the lens and the film. You can see objects clearly at various distances because the shape of your eye's lens is changed automatically by small muscles.

Directly behind the lens, on the retina, is a small, notched area called the fovea. This area contains only a high concentration of cone receptors. When you look directly at an object, the image is focused mainly on the fovea. The cones, however, do not

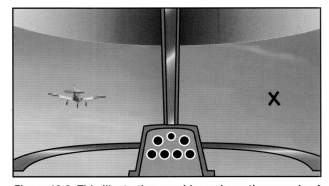

Figure 13-2. This illustration provides a dramatic example of the eye's blind spot. Cover your right eye and hold this page at arm's length. Focus your left eye on the X in the right side of the visual, and notice what happens to the aircraft as you slowly bring the page closer to your eye.

function well in darkness, which explains why you cannot see color as vividly at night as you can during the day. [Figure 13-3]

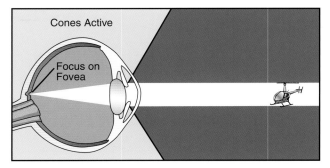

Figure 13-3. The best vision in daylight is obtained by looking directly at the object. This focuses the image on the fovea, where detail is best seen.

RODS

The rods are our dim light and night receptors and are concentrated outside the fovea area. The number of rods increases as the distance from the fovea increases. Rods sense images only in black and white. Because the rods are not located directly behind the pupil, they are responsible for much of our peripheral vision. Images that move are perceived more easily by the rod areas than by the cones in the fovea. If you have ever seen something move out of the corner of your eye, it was most likely detected by your rod receptors.

Since the cones do not function well in the dark, you may not be able to see an object if you look directly at it. The concentration of cones in the fovea can make a night blindspot at the center of your vision. To see an object clearly, you must expose the rods to the image. This is accomplished by looking 5° to 10° off center of the object you want to see. You can try out this effect on a dim light in a darkened room. When you look directly at the light, it dims or disappears altogether. If you look slightly off center, it becomes clearer and brighter. [Figure 13-4]

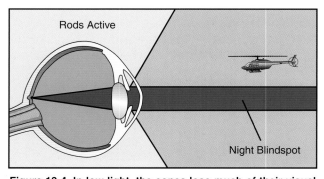

Figure 13-4. In low light, the cones lose much of their visual acuity, while rods become more receptive. The eye sacrifices sharpness for sensitivity. Your ability to see an object directly in front of you is reduced, and you lose much of your depth perception, as well as your judgment of size.

How well you see at night is determined by the rods in your eyes, as well as the amount of light allowed into your eyes. The wider the pupil is open at night, the better your night vision becomes.

NIGHT VISION

The cones in your eyes adapt quite rapidly to changes in light intensities, but the rods do not. If you have ever walked from bright sunlight into a dark movie theater, you have experienced this dark adaptation period. The rods can take approximately 30 minutes to fully adapt to the dark. A bright light, however, can completely destroy your night adaptation and severely restrict your visual acuity.

There are several things you can do to keep your eyes adapted to the dark. The first is obvious; avoid bright lights before and during the flight. For 30 minutes before a night flight, avoid any bright light sources, such as headlights, landing lights, strobe lights, or flashlights. If you encounter a bright light, close one eye to keep it light sensitive. This allows you to see again once the light is gone. Light sensitivity also can be gained by using sunglasses if you will be flying from daylight into an area of increasing darkness.

Red cockpit lighting also helps preserve your night vision, but red light severely distorts some colors, and completely washes out the color red. This makes reading an aeronautical chart difficult. A dim white light or carefully directed flashlight can enhance your night reading ability. While flying at night, keep the instrument panel and interior lights turned up no higher than necessary. This helps you see outside visual references more easily. If your eyes become blurry, blinking more frequently often helps.

Your diet and general physical health have an impact on how well you can see in the dark. Deficiencies in vitamins A and C have been shown to reduce night acuity. Other factors, such as carbon monoxide poisoning, smoking, alcohol, certain drugs, and a lack of oxygen also can greatly decrease your night vision.

NIGHT SCANNING

Good night visual acuity is needed for collision avoidance. Night scanning, like day scanning, uses a series of short, regularly spaced eye movements in 10° sectors. Unlike day scanning, however, off-center viewing is used to focus objects on the rods rather than the fovea blindspot. When you look at an object, avoid staring at it too long. If you stare at an object without moving your eyes, the retina becomes accustomed to the light intensity and the image begins to fade. To keep it clearly visible, new areas in the retina must be exposed to the image. Small, circular eye movements help eliminate the fading. You also need to move your eyes more slowly from sector to sector than during the day to prevent blurring.

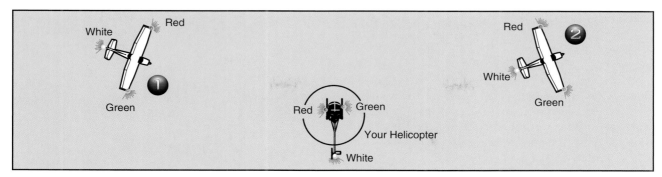

Figure 13-5. By interpreting the position lights on other aircraft, you can determine whether the aircraft is flying away from you or is on a collision course. If you see a red position light to the right of a green light, such as shown by aircraft number 1, it is flying toward you. You should watch this aircraft closely and be ready to change course. Aircraft number 2, on the other hand, is flying away from you, as indicated by the white position light.

AIRCRAFT LIGHTING

In order to see other aircraft more clearly, regulations require that all aircraft operating during the night hours have special lights and equipment. The requirements for operating at night are found in Title 14 of the Code of Federal Regulations (14 CFR) part 91. In addition to aircraft lighting, the regulations also provide a definition of nighttime, currency requirements, fuel reserves, and necessary electrical systems.

Position lights enable you to locate another aircraft, as well as help you determine its direction of flight. The approved aircraft lights for night operations are a green light on the right cabin side or wingtip, a red light on the left cabin side or wingtip, and a white position light on the tail. In addition, flashing aviation red or white anticollision lights are required for night flights. These flashing lights can be in a number of locations, but are most commonly found on the top and bottom of the cabin. [Figure 13-5]

VISUAL ILLUSIONS

There are many different types of visual illusions that you can experience at any time, day or night. The next few paragraphs cover some of the illusions that commonly occur at night.

AUTOKINESIS

Autokinesis is caused by staring at a single point of light against a dark background, such as a ground light or bright star, for more than a few seconds. After a few moments, the light appears to move on its own. To prevent this illusion, you should focus your eyes on objects at varying distances and not fixate on one target, as well as maintain a normal scan pattern.

NIGHT MYOPIA

Another problem associated with night flying is night myopia, or night-induced nearsightedness. With nothing to focus on, your eyes automatically focus on a point just slightly ahead of your aircraft. Searching out and focusing on distant light sources, no matter how dim, helps prevent the onset of night myopia.

FALSE HORIZON

A false horizon can occur when the natural horizon is obscured or not readily apparent. It can be generated by confusing bright stars and city lights. [Figure 13-6] It can also occur while you are flying toward the shore of an ocean or a large lake. Because of the relative darkness of the water, the lights along the shoreline can be mistaken for the stars in the sky. [Figure 13-7]

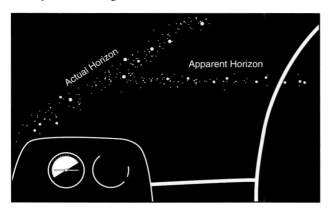

Figure 13-6. You can place your helicopter in an extremely dangerous flight attitude if you align the helicopter with the wrong lights. Here, the helicopter is aligned with a road and not the horizon.

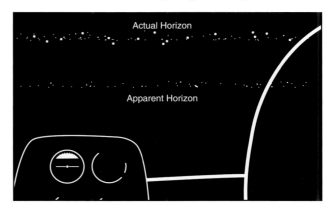

Figure 13-7. In this illusion, the shoreline is mistaken for the horizon. In an attempt to correct for the apparent nose-high attitude, a pilot may lower the collective and attempt to fly "beneath the shore."

LANDING ILLUSIONS

Landing illusions occur in many forms. Above feature-less terrain at night, there is a natural tendency to fly a lower-than-normal approach. Elements that cause any type of visual obscuration, such as rain, haze, or a dark runway environment also can cause low approaches. Bright lights, steep surrounding terrain, and a wide runway can produce the illusion of being too low, with a tendency to fly a higher-than-normal approach.

NIGHT FLIGHT

The night flying environment and the techniques you use when flying at night, depend on outside conditions. Flying on a bright, clear, moonlit evening when the visibility is good and the wind is calm, is not much different from flying during the day. However, if you are flying on an overcast night over a sparsely populated area, with little or no outside lights from the ground, the situation is quite different. Visibility is restricted so you have to be more alert in steering clear of obstructions and low clouds. Your options are also limited in the event of an emergency, as it is more difficult to find a place to land and determine wind direction and speed. At night, you have to rely more heavily on the aircraft systems, such as lights, flight instruments, and navigation equipment. *As a precaution, if the visibility is limited or outside references are inadequate, you should strongly consider delaying the flight until conditions improve, unless you have received training in instrument flight and your helicopter has the appropriate instrumentation and equipment.*

PREFLIGHT

The preflight inspection is performed in the usual manner, except it should be done in a well lit area or with a flashlight. Careful attention must be paid to the aircraft electrical system. In helicopters equipped with fuses, a spare set is required by regulation, and common sense, so make sure they are onboard. If the helicopter is equipped with circuit breakers, check to see that they are not tripped. A tripped circuit breaker may be an indication of an equipment malfunction. Reset it and check the associated equipment for proper operation.

Check all the interior lights, especially the instrument and panel lights. The panel lighting can usually be controlled with a rheostat or dimmer switch, allowing you to adjust the intensity. If the lights are too bright, a glare may reflect off the windshield creating a distraction. Always carry a flashlight with fresh batteries to provide an alternate source of light if the interior lights malfunction.

All aircraft operating between sunset and sunrise are required to have operable navigation lights. Turn these lights on during the preflight to inspect them visually for proper operation. Between sunset and sunrise, theses lights must be on any time the engine is running.

All recently manufactured aircraft certified for night flight, must have an anticollision light that makes the aircraft more visible to other pilots. This light is either a red or white flashing light and may be in the form of a rotating beacon or a strobe. While anticollision lights are required for night VFR flights, they may be turned off any time they create a distraction for the pilot.

One of the first steps in preparation for night flight is becoming thoroughly familiar with the helicopter's cockpit, instrumentation and control layout. It is recommended that you practice locating each instrument, control, and switch, both with and without cabin lights. Since the markings on some switches and circuit breaker panels may be hard to read at night, you should assure yourself that you are able to locate and use these devices, and read the markings in poor light conditions. Before you start the engine, make sure all necessary equipment and supplies needed for the flight, such as charts, notepads, and flashlights, are accessible and ready for use.

ENGINE STARTING AND ROTOR ENGAGEMENT

Use extra caution when starting the engine and engaging the rotors, especially in dark areas with little or no outside lights. In addition to the usual call of "clear," turn on the position and anticollision lights. If conditions permit, you might also want to turn the landing light on momentarily to help warn others that you are about to start the engine and engage the rotors.

TAXI TECHNIQUE

Landing lights usually cast a beam that is narrow and concentrated ahead of the helicopter, so illumination to the side is minimal. Therefore, you should slow your taxi at night, especially in congested ramp and parking areas. Some helicopters have a hover light in addition to a landing light, which illuminates a larger area under the helicopter.

When operating at an unfamiliar airport at night, you should ask for instructions or advice concerning local conditions, so as to avoid taxiing into areas of construction, or unlighted, unmarked obstructions. Ground controllers or UNICOM operators are usually cooperative in furnishing you with this type of information.

TAKEOFF

Before takeoff, make sure that you have a clear, unobstructed takeoff path. At airports, you may accomplish this by taking off over a runway or taxiway, however, if you are operating off-airport, you must pay more attention to the surroundings. Obstructions may also be difficult to see if you are taking off from an unlighted area. Once you have chosen a suitable takeoff path, select a point down the takeoff path to use for directional reference. During a night takeoff, you may notice a lack of reliable outside visual references after you are airborne. This is particularly true at small airports and off-airport landing sites located in sparsely populated areas. To compensate for the lack of outside references, use the available flight instruments as an aid. Check the altimeter and the airspeed indicator to verify the proper climb attitude. An attitude indicator, if installed, can enhance your attitude reference.

The first 500 feet of altitude after takeoff is considered to be the most critical period in transitioning from the comparatively well-lighted airport or heliport into what sometimes appears to be total darkness. A takeoff at night is usually an "altitude over airspeed" maneuver, meaning you will most likely perform a nearly maximum performance takeoff. This improves the chances for obstacle clearance and enhances safety. When performing this maneuver, be sure to avoid the cross-hatched or shaded areas of the height-velocity diagram.

EN ROUTE PROCEDURES
In order to provide a higher margin of safety, it is recommended that you select a cruising altitude somewhat higher than normal. There are several reasons for this. First, a higher altitude gives you more clearance between obstacles, especially those that are difficult to see at night, such as high tension wires and unlighted towers. Secondly, in the event of an engine failure, you have more time to set up for a landing and the gliding distance is greater giving you more options in making a safe landing. Thirdly, radio reception is improved, particularly if you are using radio aids for navigation.

During your preflight planning, it is recommended that you select a route of flight that keeps you within reach of an airport, or any safe landing site, as much of the time as possible. It is also recommended that you fly as close as possible to a populated or lighted area such as a highway or town. Not only does this offer more options in the event of an emergency, but also makes navigation a lot easier. A course comprised of a series of slight zig-zags to stay close to suitable landing sites and well lighted areas, only adds a little more time and distance to an otherwise straight course.

In the event that you have to make a forced landing at night, use the same procedure recommended for daytime emergency landings. If available, turn on the landing light during the final descent to help in avoiding obstacles along your approach path.

COLLISION AVOIDANCE AT NIGHT
At night, the outside visual references are greatly reduced especially when flying over a sparsely populated area with little or no lights. The result is that you tend to focus on a single point or instrument, making you less aware of the other traffic around. You must make a special effort to devote enough time to scan for traffic. You can determine another aircraft's direction of flight by interpreting the position and anticollision lights.

APPROACH AND LANDING
Night approaches and landings do have some advantages over daytime approaches, as the air is generally smoother and the disruptive effects of turbulence and excessive crosswinds are often absent. However, there are a few special considerations and techniques that apply to approaches at night. For example, when landing at night, especially at an unfamiliar airport, make the approach to a lighted runway and then use the taxiways to avoid unlighted obstructions or equipment.

Carefully controlled studies have revealed that pilots have a tendency to make lower approaches at night than during the day. This is potentially dangerous as you have a greater chance of hitting an obstacle, such as an overhead wire or fence, which are difficult to see. It is good practice to make steeper approaches at night, thus increasing any obstacle clearance. Monitor your altitude and rate of descent using the altimeter.

Another tendency is to focus too much on the landing area and not pay enough attention to airspeed. If too much airspeed is lost, a settling-with-power condition may result. Maintain the proper attitude during the approach, and make sure you keep some forward airspeed and movement until close to the ground. Outside visual reference for airspeed and rate of closure may not be available, especially when landing in an unlighted area, so pay special attention to the airspeed indicator

Although the landing light is a helpful aid when making night approaches, there is an inherent disadvantage. The portion of the landing area illuminated by the landing light seems higher than the dark area surrounding it. This effect can cause you to terminate the approach at too high an altitude, resulting in a settling-with-power condition and a hard landing.

Aeronautical decision making (ADM) is a systematic approach to the mental process used by pilots to consistently determine the best course of action in response to a given set of circumstances. The importance of learning effective ADM skills cannot be overemphasized. While progress is continually being made in the advancement of pilot training methods, aircraft equipment and systems, and services for pilots, accidents still occur. Despite all the changes in technology to improve flight safety, one factor remains the same—the human factor. It is estimated that approximately 65 percent of the total rotorcraft accidents are **human factors** related.

Historically, the term "pilot error" has been used to describe the causes of these accidents. Pilot error means that an action or decision made by the pilot was the cause of, or a contributing factor that lead to, the accident. This definition also includes the pilot's failure to make a decision or take action. From a broader perspective, the phrase "human factors related" more aptly describes these accidents since it is usually not a single decision that leads to an accident, but a chain of events triggered by a number of factors.

The poor judgment chain, sometimes referred to as the "error chain," is a term used to describe this concept of contributing factors in a human factors related accident. Breaking one link in the chain normally is all that is necessary to change the outcome of the sequence of events. The following is an example of the type of scenario illustrating the poor judgment chain.

A helicopter pilot, with limited experience flying in adverse weather, wants to be back at his home airport in time to attend an important social affair. He is already 30 minutes late. Therefore, he decides not to refuel his helicopter, since he should get back home with at least 20 minutes of reserve. In addition, in spite of his inexperience, he decides to fly through an area of possible thunderstorms in order to get back just before dark. Arriving in the thunderstorm area, he encounters lightning, turbulence, and heavy clouds. Night is approaching, and the thick cloud cover makes it very dark. With his limited fuel supply, he is not able to circumnavigate the thunderstorms. In the darkness and turbulence, the pilot becomes spatially disoriented while attempting to continue flying with visual reference to the ground instead of using what instruments he has to make a 180° turn. In the ensuing crash, the pilot is seriously injured and the helicopter completely destroyed.

By discussing the events that led to this accident, we can understand how a series of judgmental errors contributed to the final outcome of this flight. For example, one of the first elements that affected the pilot's flight was a decision regarding the weather. The pilot knew there were going to be thunderstorms in the area, but he had flown near thunderstorms before and never had an accident.

Next, he let his desire to arrive at his destination on time override his concern for a safe flight. For one thing, in order to save time, he did not refuel the helicopter, which might have allowed him the opportunity to circumnavigate the bad weather. Then he overestimated his flying abilities and decided to use a route that took him through a potential area of thunderstorm activity. Next, the pilot pressed on into obviously deteriorating conditions instead of changing course or landing prior to his destination.

On numerous occasions during the flight, the pilot could have made effective decisions that may have prevented this accident. However, as the chain of events unfolded, each poor decision left him with fewer and fewer options. Making sound decisions is the key to preventing accidents. Traditional pilot training has

Human Factors—The study of how people interact with their environments. In the case of general aviation, it is the study of how pilot performance is influenced by such issues as the design of cockpits, the function of the organs of the body, the effects of emotions, and the interaction and communication with the other participants of the aviation community, such as other crew members and air traffic control personnel.

emphasized flying skills, knowledge of the aircraft, and familiarity with regulations. ADM training focuses on the decision-making process and the factors that affect a pilot's ability to make effective choices.

ORIGINS OF ADM TRAINING

The airlines developed some of the first training programs that focused on improving aeronautical decision making. Human factors-related accidents motivated the airline industry to implement crew resource management (CRM) training for flight crews. The focus of CRM programs is the effective use of all available resources; human resources, hardware, and information. Human resources include all groups routinely working with the cockpit crew (or pilot) who are involved in decisions that are required to operate a flight safely. These groups include, but are not limited to: ground personnel, dispatchers, cabin crewmembers, maintenance personnel, external-load riggers, and air traffic controllers. Although the CRM concept originated as airlines developed ways of facilitating crew cooperation to improve decision making in the cockpit, CRM principles, such as workload management, situational awareness, communication, the leadership role of the captain, and crewmember coordination have direct application to the general aviation cockpit. This also includes single pilot operations since pilots of small aircraft, as well as crews of larger aircraft, must make effective use of all available resources—human resources, hardware, and information. You can also refer to AC 60-22, *Aeronautical Decision Making*, which provides background references, definitions, and other pertinent information about ADM training in the general aviation environment. [Figure 14-1]

DEFINITIONS
ADM is a systematic approach to the mental process used by pilots to consistently determine the best course of action in response to a given set of circumstances.
ATTITUDE is a personal motivational predisposition to respond to persons, situations, or events in a given manner that can, nevertheless, be changed or modified through training as sort of a mental shortcut to decision making.
ATTITUDE MANAGEMENT is the ability to recognize hazardous attitudes in oneself and the willingness to modify them as necessary through the application of an appropriate antidote thought.
CREW RESOURCE MANAGEMENT (CRM) is the application of team management concepts in the flight deck environment. It was initially known as cockpit resource management, but as CRM programs evolved to include cabin crews, maintenance personnel, and others, the phrase crew resource management was adopted. This includes single pilots, as in most general aviation aircraft. Pilots of small aircraft, as well as crews of larger aircraft, must make effective use of all available resources; human resources, hardware, and information. A current definition includes all groups routinely working with the cockpit crew who are involved in decisions required to operate a flight safely. These groups include, but are not limited to: pilots, dispatchers, cabin crewmembers, maintenance personnel, and air traffic controllers. CRM is one way of addressing the challenge of optimizing the human/machine interface and accompanying interpersonal activities.
HEADWORK is required to accomplish a conscious, rational thought process when making decisions. Good decision making involves risk identification and assessment, information processing, and problem solving.
JUDGMENT is the mental process of recognizing and analyzing all pertinent information in a particular situation, a rational evaluation of alternative actions in response to it, and a timely decision on which action to take.
PERSONALITY is the embodiment of personal traits and characteristics of an individual that are set at a very early age and extremely resistant to change.
POOR JUDGMENT CHAIN is a series of mistakes that may lead to an accident or incident. Two basic principles generally associated with the creation of a poor judgment chain are: (1) One bad decision often leads to another; and (2) as a string of bad decisions grows, it reduces the number of subsequent alternatives for continued safe flight. ADM is intended to break the poor judgment chain before it can cause an accident or incident.
RISK ELEMENTS IN ADM take into consideration the four fundamental risk elements: the pilot, the aircraft, the environment, and the type of operation that comprise any given aviation situation.
RISK MANAGEMENT is the part of the decision making process which relies on situational awareness, problem recognition, and good judgment to reduce risks associated with each flight.
SITUATIONAL AWARENESS is the accurate perception and understanding of all the factors and conditions within the four fundamental risk elements that affect safety before, during, and after the flight.
SKILLS and PROCEDURES are the procedural, psychomotor, and perceptual skills used to control a specific aircraft or its systems. They are the airmanship abilities that are gained through conventional training, are perfected, and become almost automatic through experience.
STRESS MANAGEMENT is the personal analysis of the kinds of stress experienced while flying, the application of appropriate stress assessment tools, and other coping mechanisms.

Figure 14-1. These terms are used in AC 60-22 to explain concepts used in ADM training.

THE DECISION-MAKING PROCESS

An understanding of the decision-making process provides you with a foundation for developing ADM skills. Some situations, such as engine failures, require you to respond immediately using established procedures with little time for detailed analysis. Traditionally, pilots have been well trained to react to emergencies, but are not as well prepared to make decisions that require a more reflective response. Typically during a flight, you have time to examine any changes that occur, gather information, and assess risk before reaching a decision. The steps leading to this conclusion constitute the decision-making process.

DEFINING THE PROBLEM

Problem definition is the first step in the decision-making process. Defining the problem begins with recognizing that a change has occurred or that an expected change did not occur. A problem is perceived first by the senses, then is distinguished through insight and experience. These same abilities, as well as an objective analysis of all available information, are used to determine the exact nature and severity of the problem.

While doing a hover check after picking up fire fighters at the bottom of a canyon, you realize that you are only 20 pounds under maximum gross weight. What you failed to realize is that they had stowed some of their heaviest gear in the baggage compartment, which shifted the CG slightly behind the aft limits. Since weight and balance had never created any problems for you in the past, you did not bother to calculate CG and power required. You did, however, try to estimate it by remembering the figures from earlier in the morning at the base camp. At a 5,000 foot density altitude and maximum gross weight, the performance charts indicated you had plenty of excess power. Unfortunately, the temperature was 93°F and the pressure altitude at the pick up point was 6,200 feet (DA = 9,600 feet). Since there was enough power for the hover check, you felt there was sufficient power to take off.

Even though the helicopter accelerated slowly during the takeoff, the distance between the helicopter and the ground continued to increase. However, when you attempted to establish the best rate of climb speed, the nose wanted to pitch up to a higher than normal attitude, and you noticed that the helicopter was not gaining enough altitude in relation to the canyon wall a couple hundred yards ahead.

CHOOSING A COURSE OF ACTION

After the problem has been identified, you must evaluate the need to react to it and determine the actions that need to be taken to resolve the situation in the time available. The expected outcome of each possible action should be considered and the risks assessed before you decide on a response to the situation.

Your first thought was to pull up on the collective and yank back on the cyclic, but after weighing the consequences of possibly losing rotor r.p.m. and not being able to maintain the climb rate sufficiently enough to clear the canyon wall, which is now only a hundred yards away, you realize that your only course is to try to turn back to the landing zone on the canyon floor.

IMPLEMENTING THE DECISION AND EVALUATING THE OUTCOME

Although a decision may be reached and a course of action implemented, the decision-making process is not complete. It is important to think ahead and determine how the decision could affect other phases of the flight. As the flight progresses, you must continue to evaluate the outcome of the decision to ensure that it is producing the desired result.

As you make your turn to the downwind, the airspeed drops nearly to zero, and the helicopter becomes very difficult to control. At this point, you must increase airspeed in order to maintain translational lift, but since the CG is aft of limits, you need to apply more forward cyclic than usual. As you approach the landing zone with a high rate of descent, you realize that you are in a potential settling-with-power situation if you try to trade airspeed for altitude and lose ETL. Therefore, you will probably not be able to terminate the approach in a hover. You decide to make as shallow of an approach as possible and perform a run-on landing.

The decision making process normally consists of several steps before you choose a course of action. To help you remember the elements of the decision-making process, a six-step model has been developed using the acronym "DECIDE." [Figure 14-2]

DECIDE MODEL
Detect the fact that a change has occurred.
Estimate the need to counter or react to the change.
Choose a desirable outcome for the success of the flight.
Identify actions which could successfully control the change.
Do the necessary action to adapt to the change.
Evaluate the effect of the action.

Figure 14-2. The DECIDE model can provide a framework for effective decision making.

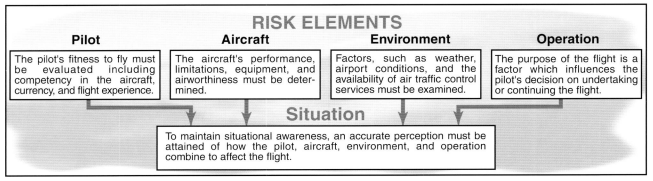

RISK ELEMENTS

Pilot	Aircraft	Environment	Operation
The pilot's fitness to fly must be evaluated including competency in the aircraft, currency, and flight experience.	The aircraft's performance, limitations, equipment, and airworthiness must be determined.	Factors, such as weather, airport conditions, and the availability of air traffic control services must be examined.	The purpose of the flight is a factor which influences the pilot's decision on undertaking or continuing the flight.

Situation

To maintain situational awareness, an accurate perception must be attained of how the pilot, aircraft, environment, and operation combine to affect the flight.

Figure 14-3. When situationally aware, you have an overview of the total operation and are not fixated on one perceived significant factor.

RISK MANAGEMENT

During each flight, decisions must be made regarding events that involve interactions between the four **risk elements**—the pilot in command, the aircraft, the environment, and the operation. The decision-making process involves an evaluation of each of these risk elements to achieve an accurate perception of the flight situation. [Figure 14-3]

One of the most important decisions that a pilot in command must make is the go/no-go decision. Evaluating each of these risk elements can help you decide whether a flight should be conducted or continued. Let us evaluate the four risk elements and how they affect our decision making regarding the following situations.

Pilot—As a pilot, you must continually make decisions about your own competency, condition of health, mental and emotional state, level of fatigue, and many other variables. For example, you are called early in the morning to make a long flight. You have had only a few hours of sleep, and are concerned that the congestion you feel could be the onset of a cold. Are you safe to fly?

Aircraft—You will frequently base decisions on your evaluations of the aircraft, such as its powerplant, performance, equipment, fuel state, or airworthiness. Picture yourself in this situation: you are en route to an oil rig an hour's flight from shore, and you have just passed the shoreline. Then you notice the oil temperature at the high end of the caution range. Should you continue out to sea, or return to the nearest suitable heliport/airport?

Environment—This encompasses many elements not pilot or aircraft related. It can include such factors as weather, air traffic control, navaids, terrain, takeoff and landing areas, and surrounding obstacles. Weather is one element that can change drastically over time and distance. Imagine you are ferrying a helicopter cross country and encounter unexpected low clouds and rain in an area of rising terrain. Do you try to stay under them and "scud run," or turn around, stay in the clear, and obtain current weather information?

Operation—The interaction between you as the pilot, your aircraft, and the environment is greatly influenced by the purpose of each flight operation. You must evaluate the three previous areas to decide on the desirability of undertaking or continuing the flight as planned. It is worth asking yourself why the flight is being made, how critical is it to maintain the schedule, and is the trip worth the risks? For instance, you are tasked to take some technicians into rugged mountains for a routine survey, and the weather is marginal. Would it be preferable to wait for better conditions to ensure a safe flight? How would the priorities change if you were tasked to search for cross-country skiers who had become lost in deep snow and radioed for help?

ASSESSING RISK

Examining **NTSB** reports and other accident research can help you to assess risk more effectively. For example, the accident rate decreases by nearly 50 percent once a pilot obtains 100 hours, and continues to decrease until the 1,000 hour level. The data suggest that for the first 500 hours, pilots flying VFR at night should establish higher personal limitations than are required by the regulations and, if applicable, apply instrument flying skills in this environment. [Figure 14-4]

Studies also indicate the types of flight activities that are most likely to result in the most serious accidents. The majority of fatal general aviation accident causes fall under the categories of maneuvering flight, approaches, takeoff/initial climb, and weather. Delving deeper into accident statistics can provide some important details that can help you to understand the risks involved with specific flying situations. For example, maneuvering flight is one of the largest single produc-

Risk Elements—The four components of a flight that make up the overall situation.

NTSB—National Transportation Safety Board.

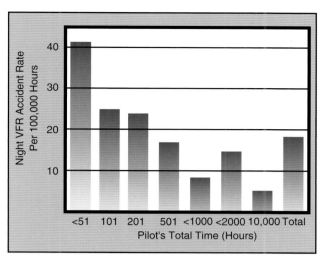

Figure 14-4. Statistical data can identify operations that have more risk.

ers of fatal accidents. Fatal accidents, which occur during approach, often happen at night or in IFR conditions. Takeoff/initial climb accidents frequently are due to the pilot's lack of awareness of the effects of density altitude on aircraft performance or other improper takeoff planning resulting in loss of control during, or shortly after takeoff. The majority of weather-related accidents occur after attempted VFR flight into IFR conditions.

FACTORS AFFECTING DECISION MAKING

It is important to point out the fact that being familiar with the decision-making process does not ensure that you will have the good judgment to be a safe pilot. The ability to make effective decisions as pilot in command depends on a number of factors. Some circumstances, such as the time available to make a decision, may be beyond your control. However, you can learn to recognize those factors that can be managed, and learn skills to improve decision-making ability and judgment.

PILOT SELF-ASSESSMENT

The pilot in command of an aircraft is directly responsible for, and is the final authority as to, the operation of that aircraft. In order to effectively exercise that responsibility and make effective decisions regarding the outcome of a flight, you must have an understanding of your limitations. Your performance during a flight is affected by many factors, such as health, recency of experience, knowledge, skill level, and attitude.

Exercising good judgment begins prior to taking the controls of an aircraft. Often, pilots thoroughly check their aircraft to determine airworthiness, yet do not evaluate their own fitness for flight. Just as a checklist

is used when preflighting an aircraft, a personal checklist based on such factors as experience, currency, and comfort level can help determine if you are prepared for a particular flight. Specifying when refresher training should be accomplished and designating weather minimums, which may be higher than those listed in Title 14 of the Code of Federal Regulations (14 CFR) part 91, are elements that may be included on a personal checklist. In addition to a review of personal limitations, you should use the I'M SAFE Checklist to further evaluate your fitness for flight. [Figure 14-5]

Figure 14-5. Prior to flight, you should assess your fitness, just as you evaluate the aircraft's airworthiness.

RECOGNIZING HAZARDOUS ATTITUDES

Being fit to fly depends on more than just your physical condition and recency of experience. For example, attitude affects the quality of your decisions. Attitude can be defined as a personal motivational predisposition to respond to persons, situations, or events in a given manner. Studies have identified five hazardous attitudes that can interfere with your ability to make sound decisions and exercise authority properly. [Figure 14-6]

Hazardous attitudes can lead to poor decision making and actions that involve unnecessary risk. You must examine your decisions carefully to ensure that your choices have not been influenced by hazardous attitudes, and you must be familiar with positive alternatives to counteract the hazardous attitudes. These substitute attitudes are referred to as antidotes. During a flight operation, it is important to be able to recognize

THE FIVE HAZARDOUS ATTITUDES	
1. Anti-Authority: "Don't tell me."	This attitude is found in people who do not like anyone telling them what to do. In a sense, they are saying, "No one can tell me what to do." They may be resentful of having someone tell them what to do, or may regard rules, regulations, and procedures as silly or unnecessary. However, it is always your prerogative to question authority if you feel it is in error.
2. Impulsivity: "Do it quickly."	This is the attitude of people who frequently feel the need to do something, anything, immediately. They do not stop to think about what they are about to do; they do not select the best alternative, and they do the first thing that comes to mind.
3. Invulnerability: "It won't happen to me."	Many people feel that accidents happen to others, but never to them. They know accidents can happen, and they know that anyone can be affected. They never really feel or believe that they will be personally involved. Pilots who think this way are more likely to take chances and increase risk.
4. Macho: "I can do it."	Pilots who are always trying to prove that they are better than anyone else are thinking, "I can do it —I'll show them." Pilots with this type of attitude will try to prove themselves by taking risks in order to impress others. While this pattern is thought to be a male characteristic, women are equally susceptible.
5. Resignation: "What's the use?"	Pilots who think, "What's the use?" do not see themselves as being able to make a great deal of difference in what happens to them. When things go well, the pilot is apt to think that it is good luck. When things go badly, the pilot may feel that someone is out to get me, or attribute it to bad luck. The pilot will leave the action to others, for better or worse. Sometimes, such pilots will even go along with unreasonable requests just to be a "nice guy."

Figure 14-6. You should examine your decisions carefully to ensure that your choices have not been influenced by a hazardous attitude.

HAZARDOUS ATTITUDES	ANTIDOTES
Macho—Brenda often brags to her friends about her skills as a pilot and wants to impress them with her abilities. During her third solo flight she decides to take a friend for a helicopter ride.	**Taking chances is foolish.**
Anti-authority—In the air she thinks "It's great to be up here without an instructor criticizing everything I do. His do-it-by-the-book attitude takes all of the fun out of flying."	**Follow the rules. They are usually right.**
Invulnerability—As she nears her friends farm, she remembers that it is about eight miles from the closest airport. She thinks, "I'll land in the pasture behind the barn at Sarah's farm. It won't be dangerous at all... the pasture is fenced and mowed and no animals are in the way. It's no more dangerous than landing at a heliport."	**It could happen to me.**
Impulsivity—After a short look, Brenda initiates an approach to her friend's pasture. Not realizing that she is landing with a tail wind, she makes a hard landing in the pasture and nearly hits the fence with the tail rotor before she gets the helicopter stopped.	**Not so fast. Think first.**
Resignation—A policeman pulls up to investigate what he believes to be an emergency landing. As Brenda is walking from the helicopter, she is supprised that anyone observed her landing. Her first thought is "if it weren't for my bad luck, this policeman wouldn't have come along and this would have been a great afternoon."	**I'm not helpless. I can make a difference.**

Figure 14-7. You must be able to identify hazardous attitudes and apply the appropriate antidote when needed.

a hazardous attitude, correctly label the thought, and then recall its antidote. [Figure 14-7]

STRESS MANAGEMENT

Everyone is stressed to some degree all the time. A certain amount of stress is good since it keeps a person alert and prevents complacency. However, effects of stress are cumulative and, if not coped with adequately, they eventually add up to an intolerable burden. Performance generally increases with the onset of stress, peaks, and then begins to fall off rapidly as stress levels exceed a person's ability to cope. The ability to make effective decisions during flight can be impaired by stress. Factors, referred to as stressors, can increase a pilot's risk of error in the cockpit. [Figure 14-8]

There are several techniques to help manage the accumulation of life stresses and prevent stress overload. For example, including relaxation time in a busy schedule and maintaining a program of physical fitness can help reduce stress levels. Learning to manage time more effectively can help you avoid heavy pressures imposed by getting behind schedule and not meeting deadlines. Take an assessment of yourself to determine your capabilities and limitations and then set realistic goals. In addition, avoiding stressful situations and encounters can help you cope with stress.

USE OF RESOURCES

To make informed decisions during flight operations, you must be aware of the resources found both inside and outside the cockpit. Since useful tools and sources of information may not always be readily apparent, learning to recognize these resources is an essential part of ADM training. Resources must not only be iden-

STRESSORS

Physical Stress—Conditions associated with the environment, such as temperature and humidity extremes, noise, vibration, and lack of oxygen.

Physiological Stress—Physical conditions, such as fatigue, lack of physical fitness, sleep loss, missed meals (leading to low blood sugar levels), and illness.

Psychological Stress—Social or emotional factors, such as a death in the family, a divorce, a sick child, or a demotion at work. This type of stress may also be related to mental workload, such as analyzing a problem, navigating an aircraft, or making decisions.

Figure 14-8. The three types of stressors that can affect a pilot's performance.

tified, but you must develop the skills to evaluate whether you have the time to use a particular resource and the impact that its use will have upon the safety of flight. For example, the assistance of ATC may be very useful if you are lost. However, in an emergency situation when action needs be taken quickly, time may not be available to contact ATC immediately.

INTERNAL RESOURCES
Internal resources are found in the cockpit during flight. Since some of the most valuable internal resources are ingenuity, knowledge, and skill, you can expand cockpit resources immensely by improving these capabilities. This can be accomplished by frequently reviewing flight information publications, such as the CFRs and the AIM, as well as by pursuing additional training.

A thorough understanding of all the equipment and systems in the aircraft is necessary to fully utilize all resources. For example, advanced navigation and autopilot systems are valuable resources. However, if pilots do not fully understand how to use this equipment, or they rely on it so much that they become complacent, it can become a detriment to safe flight.

Checklists are essential cockpit resources for verifying that the aircraft instruments and systems are checked, set, and operating properly, as well as ensuring that the proper procedures are performed if there is a system malfunction or in-flight emergency. In addition, the FAA-approved rotorcraft flight manual, which is required to be carried on board the aircraft, is essential for accurate flight planning and for resolving in-flight equipment malfunctions. Other valuable cockpit resources include current aeronautical charts, and publications, such as the *Airport/Facility Directory*.

Passengers can also be a valuable resource. Passengers can help watch for traffic and may be able to provide

information in an irregular situation, especially if they are familiar with flying. A strange smell or sound may alert a passenger to a potential problem. As pilot in command, you should brief passengers before the flight to make sure that they are comfortable voicing any concerns.

EXTERNAL RESOURCES
Possibly the greatest external resources during flight are air traffic controllers and flight service specialists. ATC can help decrease pilot workload by providing traffic advisories, radar vectors, and assistance in emergency situations. Flight service stations can provide updates on weather, answer questions about airport conditions, and may offer direction-finding assistance. The services provided by ATC can be invaluable in enabling you to make informed in-flight decisions.

WORKLOAD MANAGEMENT
Effective workload management ensures that essential operations are accomplished by planning, prioritizing, and sequencing tasks to avoid work overload. As experience is gained, you learn to recognize future workload requirements and can prepare for high workload periods during times of low workload. Reviewing the appropriate chart and setting radio frequencies well in advance of when they are needed helps reduce workload as your flight nears the airport. In addition, you should listen to ATIS, ASOS, or AWOS, if available, and then monitor the tower frequency or CTAF to get a good idea of what traffic conditions to expect. Checklists should be performed well in advance so there is time to focus on traffic and ATC instructions. These procedures are especially important prior to entering a high-density traffic area, such as Class B airspace.

To manage workload, items should be prioritized. For example, during any situation, and especially in an emergency, you should remember the phrase "aviate, navigate, and communicate." This means that the first thing you should do is make sure the helicopter is under

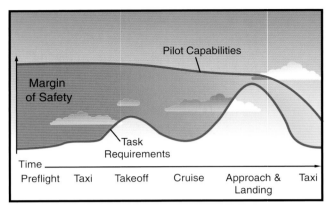

Figure 14-9. Accidents often occur when flying task requirements exceed pilot capabilities. The difference between these two factors is called the margin of safety. Note that in this idealized example, the margin of safety is minimal during the approach and landing. At this point, an emergency or distraction could overtax pilot capabilities, causing an accident.

control. Then begin flying to an acceptable landing area. Only after the first two items are assured, should you try to communicate with anyone.

Another important part of managing workload is recognizing a work overload situation. The first effect of high workload is that you begin to work faster. As workload increases, attention cannot be devoted to several tasks at one time, and you may begin to focus on one item. When you become task saturated, there is no awareness of inputs from various sources, so decisions may be made on incomplete information, and the possibility of error increases. [Figure 14-9]

When becoming overloaded, you should stop, think, slow down, and prioritize. It is important that you understand options that may be available to decrease workload. For example, tasks, such as locating an item on a chart or setting a radio frequency, may be delegated to another pilot or passenger, an autopilot, if available, may be used, or ATC may be enlisted to provide assistance.

SITUATIONAL AWARENESS

Situational awareness is the accurate perception of the operational and environmental factors that affect the aircraft, pilot, and passengers during a specific period of time. Maintaining situational awareness requires an understanding of the relative significance of these factors and their future impact on the flight. When situationally aware, you have an overview of the total operation and are not fixated on one perceived significant factor. Some of the elements inside the aircraft to be considered are the status of aircraft systems, you as the pilot, and passengers. In addition, an awareness of the environmental conditions of the flight, such as spatial orientation of the helicopter, and its relationship to terrain, traffic, weather, and airspace must be maintained.

To maintain situational awareness, all of the skills involved in aeronautical decision making are used. For example, an accurate perception of your fitness can be achieved through self-assessment and recognition of hazardous attitudes. A clear assessment of the status of navigation equipment can be obtained through workload management, and establishing a productive relationship with ATC can be accomplished by effective resource use.

OBSTACLES TO MAINTAINING SITUATIONAL AWARENESS

Fatigue, stress, and work overload can cause you to fixate on a single perceived important item rather than maintaining an overall awareness of the flight situation. A contributing factor in many accidents is a distraction that diverts the pilot's attention from monitoring the instruments or scanning outside the aircraft. Many cockpit distractions begin as a minor problem, such as a gauge that is not reading correctly, but result in accidents as the pilot diverts attention to the perceived problem and neglects to properly control the aircraft.

Complacency presents another obstacle to maintaining situational awareness. When activities become routine, you may have a tendency to relax and not put as much effort into performance. Like fatigue, complacency reduces your effectiveness in the cockpit. However, complacency is harder to recognize than fatigue, since everything is perceived to be progressing smoothly. For example, you have just dropped off another group of fire fighters for the fifth time that day. Without thinking, you hastily lift the helicopter off the ground, not realizing that one of the skids is stuck between two rocks. The result is dynamic rollover and a destroyed helicopter.

OPERATIONAL PITFALLS

There are a number of classic behavioral traps into which pilots have been known to fall. Pilots, particularly those with considerable experience, as a rule, always try to complete a flight as planned, please passengers, and meet schedules. The basic drive to meet or exceed goals can have an adverse effect on safety, and can impose an unrealistic assessment of piloting skills under stressful conditions. These tendencies ultimately may bring about practices that are dangerous and often illegal, and may lead to a mishap. You will develop awareness and learn to avoid many of these operational pitfalls through effective ADM training. [Figure 14-10]

OPERATIONAL PITFALLS

Peer Pressure—Poor decision making may be based upon an emotional response to peers, rather than evaluating a situation objectively.

Mind Set—A pilot displays mind set through an inability to recognize and cope with changes in a given situation.

Get-There-Itis—This disposition impairs pilot judgment through a fixation on the original goal or destination, combined with a disregard for any alternative course of action.

Scud Running—This occurs when a pilot tries to maintain visual contact with the terrain at low altitudes while instrument conditions exist.

Continuing Visual Flight Rules (VFR) into Instrument Conditions—Spatial disorientation or collision with ground/obstacles may occur when a pilot continues VFR into instrument conditions. This can be even more dangerous if the pilot is not instrument-rated or current.

Getting Behind the Aircraft—This pitfall can be caused by allowing events or the situation to control pilot actions. A constant state of surprise at what happens next may be exhibited when the pilot is getting behind the aircraft.

Loss of Positional or Situational Awareness—In extreme cases, when a pilot gets behind the aircraft, a loss of positional or situational awareness may result. The pilot may not know the aircraft's geographical location, or may be unable to recognize deteriorating circumstances.

Operating Without Adequate Fuel Reserves—Ignoring minimum fuel reserve requirements is generally the result of overconfidence, lack of flight planning, or disregarding applicable regulations.

Flying Outside the Envelope—The assumed high performance capability of a particular aircraft may cause a mistaken belief that it can meet the demands imposed by a pilot's overestimated flying skills.

Neglect of Flight Planning, Preflight Inspections, and Checklists—A pilot may rely on short- and long-term memory, regular flying skills, and familiar routes instead of established procedures and published checklists. This can be particularly true of experienced pilots.

Figure 14-10. All experienced pilots have fallen prey to, or have been tempted by, one or more of these tendencies in their flying careers.

CHAPTER 15

Introduction to the Gyroplane

January 9th, 1923, marked the first officially observed flight of an autogyro. The aircraft, designed by Juan de la Cierva, introduced rotor technology that made forward flight in a rotorcraft possible. Until that time, rotary-wing aircraft designers were stymied by the problem of a rolling moment that was encountered when the aircraft began to move forward. This rolling moment was the product of airflow over the rotor disc, causing an increase in lift of the advancing blade and decrease in lift of the retreating blade. Cierva's successful design, the C.4, introduced the articulated rotor, on which the blades were hinged and allowed to flap. This solution allowed the advancing blade to move upward, decreasing angle of attack and lift, while the retreating blade would swing downward, increasing angle of attack and lift. The result was balanced lift across the rotor disc regardless of airflow. This breakthrough was instrumental in the success of the modern helicopter, which was developed over 15 years later. (For more information on dissymmetry of lift, refer to Chapter 3—Aerodynamics of Flight.) On April 2, 1931, the Pitcairn PCA-2 autogyro was granted Type Certificate No. 410 and became the first rotary wing aircraft to be certified in the United States. The term "autogyro" was used to describe this type of aircraft until the FAA later designated them "gyroplanes."

By definition, the gyroplane is an aircraft that achieves lift by a free spinning rotor. Several aircraft have used the free spinning rotor to attain performance not available in the pure helicopter. The "gyrodyne" is a hybrid rotorcraft that is capable of hovering and yet cruises in autorotation. The first successful example of this type of aircraft was the British Fairy Rotodyne, certificated to the Transport Category in 1958. During the 1960s and 1970s, the popularity of gyroplanes increased with the certification of the McCulloch J-2 and Umbaugh. The latter becoming the Air & Space 18A.

There are several aircraft under development using the free spinning rotor to achieve rotary wing takeoff performance and fixed wing cruise speeds. The gyroplane offers inherent safety, simplicity of operation, and outstanding short field point-to-point capability.

TYPES OF GYROPLANES

Because the free spinning rotor does not require an antitorque device, a single rotor is the predominate configuration. Counter-rotating blades do not offer any particular advantage. The rotor system used in a gyroplane may have any number of blades, but the most popular are the two and three blade systems. Propulsion for gyroplanes may be either tractor or pusher, meaning the engine may be mounted on the front and pull the aircraft, or in the rear, pushing it through the air. The powerplant itself may be either reciprocating or turbine. Early gyroplanes were often a derivative of tractor configured airplanes with the rotor either replacing the wing or acting in conjunction with it. However, the pusher configuration is generally more maneuverable due to the placement of the rudder in the propeller slipstream, and also has the advantage of better visibility for the pilot. [Figure 15-1]

Figure 15-1. The gyroplane may have wings, be either tractor or pusher configured, and could be turbine or propeller powered. Pictured are the Pitcairn PCA-2 Autogyro (left) and the Air & Space 18A gyroplane.

When **direct control** of the rotor head was perfected, the jump takeoff gyroplane was developed. Under the proper conditions, these gyroplanes have the ability to lift off vertically and transition to forward flight. Later developments have included retaining the direct control rotor head and utilizing a wing to **unload** the rotor, which results in increased forward speed.

COMPONENTS

Although gyroplanes are designed in a variety of configurations, for the most part the basic components are the same. The minimum components required for a functional gyroplane are an airframe, a powerplant, a rotor system, tail surfaces, and landing gear. [Figure 15-2] An optional component is the wing, which is incorporated into some designs for specific performance objectives.

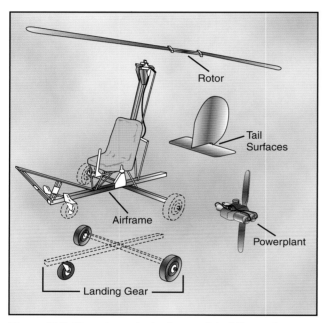

Figure 15-2. Gyroplanes typically consist of five major components. A sixth, the wing, is utilized on some designs.

AIRFRAME

The airframe provides the structure to which all other components are attached. Airframes may be welded tube, sheet metal, composite, or simply tubes bolted together. A combination of construction methods may also be employed. The airframes with the greatest strength-to-weight ratios are a carbon fiber material or the welded tube structure, which has been in use for a number of years.

POWERPLANT

The powerplant provides the thrust necessary for forward flight, and is independent of the rotor system while in flight. While on the ground, the engine may be used as a source of power to **prerotate** the rotor system. Over the many years of gyroplane development, a wide variety of engine types have been adapted to the gyroplane. Automotive, marine, ATV, and certificated aircraft engines have all been used in various gyroplane designs. Certificated gyroplanes are required to use FAA certificated engines. The cost of a new certificated aircraft engine is greater than the cost of nearly any other new engine. This added cost is the primary reason other types of engines are selected for use in amateur built gyroplanes.

ROTOR SYSTEM

The rotor system provides lift and control for the gyroplane. The fully articulated and semi-rigid teetering rotor systems are the most common. These are explained in-depth in Chapter 5—Main Rotor System. The teeter blade with hub tilt control is most common in homebuilt gyroplanes. This system may also employ a collective control to change the pitch of the rotor blades. With sufficient blade inertia and collective pitch change, jump takeoffs can be accomplished.

TAIL SURFACES

The tail surfaces provide stability and control in the pitch and yaw axes. These tail surfaces are similar to an airplane empennage and may be comprised of a fin and rudder, stabilizer and elevator. An aft mounted duct enclosing the propeller and rudder has also been used. Many gyroplanes do not incorporate a horizontal tail surface.

On some gyroplanes, especially those with an enclosed cockpit, the yaw stability is marginal due to the large fuselage side area located ahead of the center of gravity. The additional vertical tail surface necessary to compensate for this instability is difficult to achieve as the confines of the rotor tilt and high landing pitch attitude limits the available area. Some gyroplane designs incorporate multiple vertical stabilizers and rudders to add additional yaw stability.

Direct Control—The capacity for the pilot to maneuver the aircraft by tilting the rotor disc and, on some gyroplanes, affect changes in pitch to the rotor blades. These equate to cyclic and collective control, which were not available in earlier autogyros.

Unload—To reduce the component of weight supported by the rotor system.

Prerotate—Spinning a gyroplane rotor to sufficient r.p.m. prior to flight.

LANDING GEAR

The landing gear provides the mobility while on the ground and may be either conventional or tricycle. Conventional gear consists of two main wheels, and one under the tail. The tricycle configuration also uses two mains, with the third wheel under the nose. Early auto-gyros, and several models of gyroplanes, use conventional gear, while most of the later gyroplanes incorporate tricycle landing gear. As with fixed wing aircraft, the gyroplane landing gear provides the ground mobility not found in most helicopters.

WINGS

Wings may or may not comprise a component of the gyroplane. When used, they provide increased performance, increased storage capacity, and increased stability. Gyroplanes are under development with wings that are capable of almost completely unloading the rotor system and carrying the entire weight of the aircraft. This will allow rotary wing takeoff performance with fixed wing cruise speeds. [Figure 15-3]

Figure 15-3. The CarterCopter uses wings to enhance performance.

Aerodynamics of the Gyroplane

Helicopters and gyroplanes both achieve lift through the use of airfoils, and, therefore, many of the basic aerodynamic principles governing the production of lift apply to both aircraft. These concepts are explained in depth in Chapter 2—General Aerodynamics, and constitute the foundation for discussing the aerodynamics of a gyroplane.

AUTOROTATION

A fundamental difference between helicopters and gyroplanes is that in powered flight, a gyroplane rotor system operates in autorotation. This means the rotor spins freely as a result of air flowing up through the blades, rather than using engine power to turn the blades and draw air from above. [Figure 16-1] Forces are created during autorotation that keep the rotor blades turning, as well as creating lift to keep the aircraft aloft. Aerodynamically, the rotor system of a gyroplane in normal flight operates like a helicopter rotor during an engine-out forward autorotative descent.

VERTICAL AUTOROTATION

During a vertical autorotation, two basic components contribute to the relative wind striking the rotor blades. [Figure 16-2] One component, the upward flow of air through the rotor system, remains relatively constant for a given flight condition. The other component is the rotational airflow, which is the wind velocity across the blades as they spin. This component varies significantly based upon how far from the rotor hub it is measured. For example, consider a rotor disc that is 25 feet in diameter operating at 300 r.p.m. At a point one foot outboard from the rotor hub, the blades are traveling in a circle with a circumference of 6.3 feet. This equates to 31.4 feet per second (f.p.s.), or a rotational blade speed of 21 m.p.h. At the blade tips, the circumference of the circle increases to 78.5 feet. At the same operating speed of 300 r.p.m., this creates a blade tip

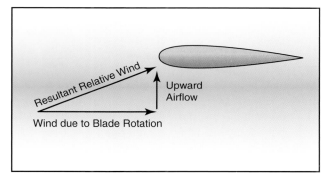

Figure 16-2. In a vertical autorotation, the wind from the rotation of the blade combines with the upward airflow to produce the resultant relative wind striking the airfoil.

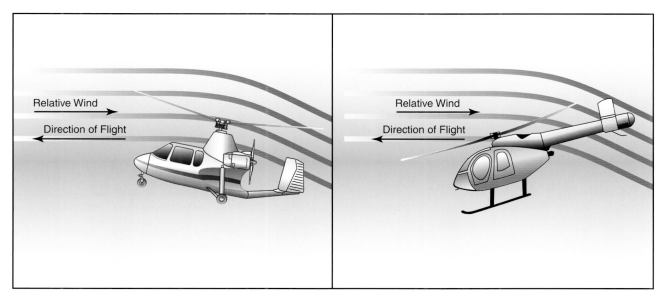

Figure 16-1. Airflow through the rotor system on a gyroplane is reversed from that on a powered helicopter. This airflow is the medium through which power is transferred from the gyroplane engine to the rotor system to keep it rotating.

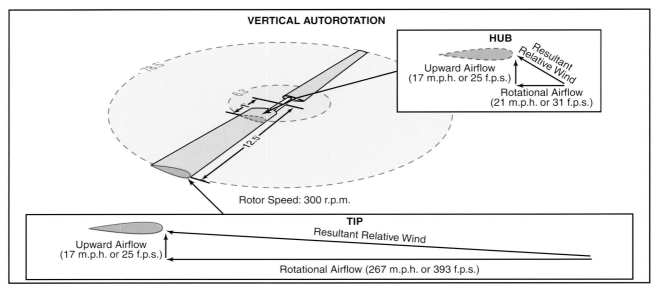

Figure 16-3. Moving outboard on the rotor blade, the rotational velocity increasingly exceeds the upward component of airflow, resulting in a higher relative wind at a lower angle of attack.

speed of 393 feet per second, or 267 m.p.h. The result is a higher total relative wind, striking the blades at a lower angle of attack. [Figure 16-3]

ROTOR DISC REGIONS

As with any airfoil, the lift that is created by rotor blades is perpendicular to the relative wind. Because the relative wind on rotor blades in autorotation shifts from a high angle of attack inboard to a lower angle of attack outboard, the lift generated has a higher forward component closer to the hub and a higher vertical component toward the blade tips. This creates distinct regions of the rotor disc that create the forces necessary for flight in autorotation. [Figure 16-4] The autorotative region, or driving region, creates a total aerodynamic force with a forward component that exceeds all rearward drag forces and keeps the blades spinning. The propeller region, or driven region, generates a total aerodynamic force with a higher vertical component that allows the gyroplane to remain aloft. Near the center of the rotor disc is a stall region where the rotational component of the relative wind is so low that the resulting angle of attack is beyond the stall limit of the airfoil. The stall region creates drag against the direction of rotation that must be overcome by the forward acting forces generated by the driving region.

AUTOROTATION IN FORWARD FLIGHT

As discussed thus far, the aerodynamics of autorotation apply to a gyroplane in a vertical descent. Because gyroplanes are normally operated in forward flight, the component of relative wind striking the rotor blades as a result of forward speed must also be considered. This component has no effect on the aerodynamic principles that cause the blades to autorotate, but causes a shift in the zones of the rotor disc.

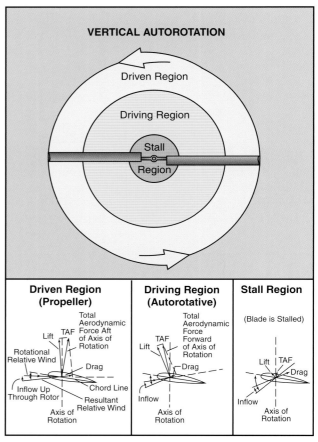

Figure 16-4. The total aerodynamic force is aft of the axis of rotation in the driven region and forward of the axis of rotation in the driving region. Drag is the major aerodynamic force in the stall region. For a complete depiction of force vectors during a vertical autorotation, refer to Chapter 3— Aerodynamics of Flight (Helicopter), Figure 3-22.

As a gyroplane moves forward through the air, the forward speed of the aircraft is effectively added to the relative wind striking the advancing blade, and sub-tracted from the relative wind striking the retreating blade. To prevent uneven lifting forces on the two sides of the rotor disc, the advancing blade teeters up, decreasing angle of attack and lift, while the retreating blade teeters down, increasing angle of attack and lift. (For a complete discussion on dissymmetry of lift, refer to Chapter 3—Aerodynamics of Flight.) The lower angles of attack on the advancing blade cause more of the blade to fall in the driven region, while higher angles of attack on the retreating blade cause more of the blade to be stalled. The result is a shift in the rotor regions toward the retreating side of the disc to a degree directly related to the forward speed of the aircraft. [Figure 16-5]

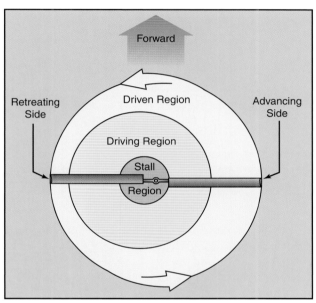

Figure 16-5. Rotor disc regions in forward autorotative flight.

REVERSE FLOW

On a rotor system in forward flight, reverse flow occurs near the rotor hub on the retreating side of the rotor disc. This is the result of the forward speed of the air-craft exceeding the rotational speed of the rotor blades. For example, two feet outboard from the rotor hub, the blades travel in a circle with a circumference of 12.6 feet. At a rotor speed of 300 r.p.m., the blade speed at the two-foot station is 42 m.p.h. If the aircraft is being operated at a forward speed of 42 m.p.h., the forward speed of the aircraft essentially negates the rotational velocity on the retreating blade at the two-foot station. Moving inboard from the two-foot station on the retreating blade, the forward speed of the aircraft increasingly exceeds the rotational velocity of the blade. This causes the airflow to actually strike the trailing edge of the rotor blade, with velocity increas-ing toward the rotor hub. [Figure 16-6] The size of the area that experiences reverse flow is dependent prima-

rily on the forward speed of the aircraft, with higher speed creating a larger region of reverse flow. To some degree, the operating speed of the rotor system also has an effect on the size of the region, with systems operat-ing at lower r.p.m. being more susceptible to reverse flow and allowing a greater portion of the blade to experience the effect.

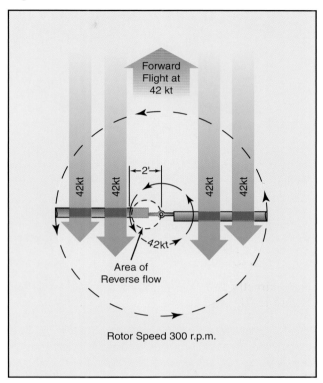

Figure 16-6. An area of reverse flow forms on the retreating blade in forward flight as a result of aircraft speed exceeding blade rotational speed.

RETREATING BLADE STALL

The retreating blade stall in a gyroplane differs from that of a helicopter in that it occurs outboard from the rotor hub at the 20 to 40 percent position rather than at the blade tip. Because the gyroplane is operating in autorotation, in forward flight there is an inherent stall region centered inboard on the retreating blade. [Refer to figure 16-5] As forward speed increases, the angle of attack on the retreating blade increases to prevent dis-symmetry of lift and the stall region moves further outboard on the retreating blade. Because the stalled portion of the rotor disc is inboard rather than near the tip, as with a helicopter, less force is created about the aircraft center of gravity. The result is that you may feel a slight increase in vibration, but you would not experi-ence a large pitch or roll tendency.

ROTOR FORCE

As with any heavier than air aircraft, the four forces acting on the gyroplane in flight are lift, weight, thrust and drag. The gyroplane derives lift from the rotor and

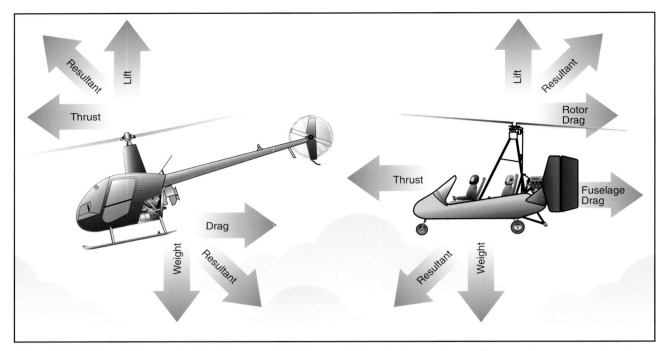

Figure 16-7. Unlike a helicopter, in forward powered flight the resultant rotor force of a gyroplane acts in a rearward direction.

thrust directly from the engine through a propeller. [Figure 16-7]

The force produced by the gyroplane rotor may be divided into two components; rotor lift and rotor drag. The component of rotor force perpendicular to the flight path is rotor lift, and the component of rotor force parallel to the flight path is rotor drag. To derive the total aircraft drag reaction, you must also add the drag of the fuselage to that of the rotor.

ROTOR LIFT
Rotor lift can most easily be visualized as the lift required to support the weight of the aircraft. When an airfoil produces lift, induced drag is produced. The most efficient angle of attack for a given airfoil produces the most lift for the least drag. However, the airfoil of a rotor blade does not operate at this efficient angle throughout the many changes that occur in each revolution. Also, the rotor system must remain in the autorotative (low) pitch range to continue turning in order to generate lift.

Some gyroplanes use small wings for creating lift when operating at higher cruise speeds. The lift provided by the wings can either supplement or entirely replace rotor lift while creating much less induced drag.

ROTOR DRAG
Total rotor drag is the summation of all the drag forces acting on the airfoil at each blade position. Each blade position contributes to the total drag according to the speed and angle of the airfoil at that position. As the

rotor blades turn, rapid changes occur on the airfoils depending on position, rotor speed, and aircraft speed. A change in the angle of attack of the rotor disc can effect a rapid and substantial change in total rotor drag.

Rotor drag can be divided into components of induced drag and profile drag. The induced drag is a product of lift, while the profile drag is a function of rotor r.p.m. Because induced drag is a result of the rotor providing lift, profile drag can be considered the drag of the rotor when it is not producing lift. To visualize profile drag, consider the drag that must be overcome to prerotate the rotor system to flight r.p.m. while the blades are producing no lift. This can be achieved with a rotor system having a symmetrical airfoil and a pitch change capability by setting the blades to a 0° angle of attack. A rotor system with an asymmetrical airfoil and a built in pitch angle, which includes most amateur-built teeter-head rotor systems, cannot be prerotated without having to overcome the induced drag created as well.

THRUST
Thrust in a gyroplane is defined as the component of total propeller force parallel to the relative wind. As with any force applied to an aircraft, thrust acts around the center of gravity. Based upon where the thrust is applied in relation to the aircraft center of gravity, a relatively small component may be perpendicular to the relative wind and can be considered to be additive to lift or weight.

In flight, the fuselage of a gyroplane essentially acts as a plumb suspended from the rotor, and as such, it is subject to **pendular action** in the same way as a helicopter. Unlike a helicopter, however, thrust is applied directly to the airframe of a gyroplane rather than being obtained through the rotor system. As a result, different forces act on a gyroplane in flight than on a helicopter. Engine torque, for example, tends to roll the fuselage in the direction opposite propeller rotation, causing it to be deflected a few degrees out of the vertical plane. [Figure 16-8] This slight "out of vertical" condition is usually negligible and not considered relevant for most flight operations.

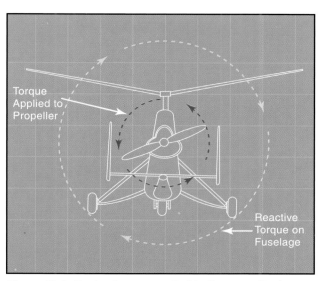

Figure 16-8. Engine torque applied to the propeller has an equal and opposite reaction on the fuselage, deflecting it a few degrees out of the vertical plane in flight.

STABILITY

Stability is designed into aircraft to reduce pilot workload and increase safety. A stable aircraft, such as a typical general aviation training airplane, requires less attention from the pilot to maintain the desired flight attitude, and will even correct itself if disturbed by a gust of wind or other outside forces. Conversely, an unstable aircraft requires constant attention to maintain control of the aircraft.

Pendular Action—The lateral or longitudinal oscillation of the fuselage due to it being suspended from the rotor system. It is similar to the action of a pendulum. Pendular action is further discussed in Chapter 3—Aerodynamics of Flight.

There are several factors that contribute to the stability of a gyroplane. One is the location of the horizontal stabilizer. Another is the location of the fuselage drag in relation to the center of gravity. A third is the inertia moment around the pitch axis, while a fourth is the relation of the propeller thrust line to the vertical location of the center of gravity (CG). However, the one that is probably the most critical is the relation of the rotor force line to the horizontal location of the center of gravity.

HORIZONTAL STABILIZER

A horizontal stabilizer helps in longitudinal stability, with its efficiency greater the further it is from the center of gravity. It is also more efficient at higher airspeeds because lift is proportional to the square of the airspeed. Since the speed of a gyroplane is not very high, manufacturers can achieve the desired stability by varying the size of the horizontal stabilizer, changing the distance it is from the center of gravity, or by placing it in the propeller slipstream.

FUSELAGE DRAG (CENTER OF PRESSURE)

If the location, where the fuselage drag or center of pressure forces are concentrated, is behind the CG, the gyroplane is considered more stable. This is especially true of yaw stability around the vertical axis. However, to achieve this condition, there must be a sufficient vertical tail surface. In addition, the gyroplane needs to have a balanced longitudinal center of pressure so there is sufficient cyclic movement to prevent the nose from tucking under or lifting, as pressure builds on the frontal area of the gyroplane as airspeed increases.

PITCH INERTIA

Without changing the overall weight and center of gravity of a gyroplane, the further weights are placed from the CG, the more stable the gyroplane. For example, if the pilot's seat could be moved forward from the CG, and the engine moved aft an amount, which keeps the center of gravity in the same location, the gyroplane becomes more stable. A tightrope walker applies this same principle when he uses a long pole to balance himself.

PROPELLER THRUST LINE

Considering just the propeller thrust line by itself, if the thrust line is above the center of gravity, the gyroplane has a tendency to pitch nose down when power is applied, and to pitch nose up when power is removed. The opposite is true when the propeller thrust line is below the CG. If the thrust line goes through the CG or

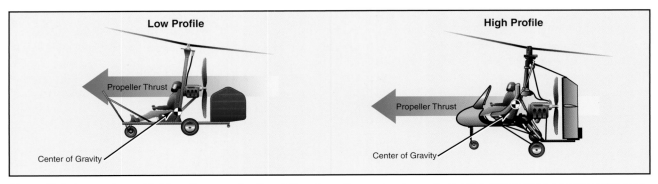

Figure 16-9. A gyroplane which has the propeller thrust line above the center of gravity is often referred to as a low profile gyroplane. One that has the propeller thrust line below or at the CG is considered a high profile gyroplane.

nearly so there is no tendency for the nose to pitch up or down. [Figure 16-9]

ROTOR FORCE

Because some gyroplanes do not have horizontal stabilizers, and the propeller thrust lines are different, gyroplane manufacturers can achieve the desired stability by placing the center of gravity in front of or behind the rotor force line. [Figure 16-10]

Suppose the CG is located behind the rotor force line in forward flight. If a gust of wind increases the angle of attack, rotor force increases. There is also an increase in the difference between the lift produced on the advancing and retreating blades. This increases the flapping angle and causes the rotor to pitch up. This pitching action increases the moment around the center of gravity, which leads to a greater increase in the angle of attack. The result is an unstable condition.

If the CG is in front of the rotor force line, a gust of wind, which increases the angle of attack, causes the rotor disc to react the same way, but now the increase in rotor force and **blade flapping** decreases the moment. This tends to decrease the angle of attack, and creates a stable condition.

TRIMMED CONDITION

As was stated earlier, manufacturers use a combination of the various stability factors to achieve a trimmed gyroplane. For example, if you have a gyroplane where the CG is below the propeller thrust line, the propeller thrust gives your aircraft a nose down pitching moment when power is applied. To compensate for this pitching moment, the CG, on this type of gyroplane, is usually located behind the rotor force line. This location produces a nose up pitching moment.

Conversely, if the CG is above the propeller thrust line, the CG is usually located ahead of the rotor force line. Of course, the location of fuselage drag, the pitch inertia, and the addition of a horizontal stabilizer can alter where the center of gravity is placed.

Blade Flapping—The upward or downward movement of the rotorblades during rotation.

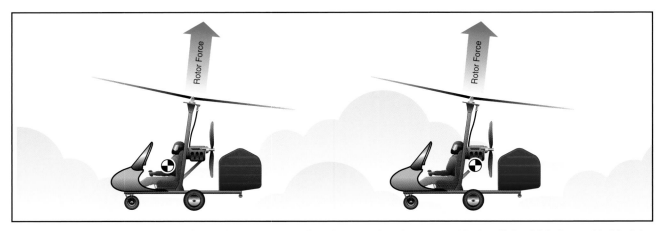

Figure 16-10. If the CG is located in front of the rotor force line, the gyroplane is more stable than if the CG is located behind the rotor force line.

Gyroplane Flight Controls

Due to rudimentary flight control systems, early gyroplanes suffered from limited maneuverability. As technology improved, greater control of the rotor system and more effective control surfaces were developed. The modern gyroplane, while continuing to maintain an element of simplicity, now enjoys a high degree of maneuverability as a result of these improvements.

CYCLIC CONTROL

The cyclic control provides the means whereby you are able to tilt the rotor system to provide the desired results. Tilting the rotor system provides all control for climbing, descending, and banking the gyroplane. The most common method to transfer stick movement to the rotor head is through push-pull tubes or flex cables. [Figure 17-1] Some gyroplanes use a direct overhead stick attachment rather than a cyclic, where a rigid control is attached to the rotor hub and descends over and in front of the pilot. [Figure 17-2] Because of the nature of the direct attachment, control inputs with this system are reversed from those used with a cyclic. Pushing forward on the control causes the rotor disc to tilt back and the gyroplane to climb, pulling back on the control initiates a descent. Bank commands are reversed in the same way.

THROTTLE

The throttle is conventional to most powerplants, and provides the means for you to increase or decrease engine power and thus, thrust. Depending on how

Figure 17-2. The direct overhead stick attachment has been used for control of the rotor disc on some gyroplanes.

the control is designed, control movement may or may not be proportional to engine power. With many gyroplane throttles, 50 percent of the control travel may equate to 80 or 90 percent of available power. This varying degree of sensitivity makes it necessary

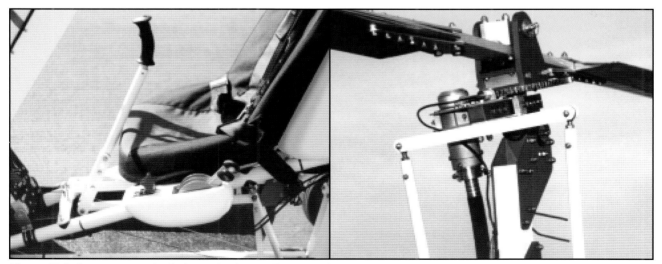

Figure 17-1. A common method of transferring cyclic control inputs to the rotor head is through the use of push-pull tubes, located outboard of the rotor mast pictured on the right.

Figure 17-3. Foot pedals provide rudder control and operation is similar to that of an airplane.

for you to become familiar with the unique throttle characteristics and engine responses for a particular gyroplane.

RUDDER

The rudder is operated by foot pedals in the cockpit and provides a means to control yaw movement of the aircraft. [Figure 17-3] On a gyroplane, this control is achieved in a manner more similar to the rudder of an airplane than to the antitorque pedals of a helicopter. The rudder is used to maintain coordinated flight, and at times may also require inputs to compensate for propeller torque. Rudder sensitivity and effectiveness are directly proportional to the velocity of airflow over the rudder surface. Consequently, many gyroplane rudders are located in the propeller slipstream and provide excellent control while the engine is developing thrust. This type of rudder configuration, however, is less effective and requires greater deflection when the engine is idled or stopped.

HORIZONTAL TAIL SURFACES

The horizontal tail surfaces on most gyroplanes are not controllable by the pilot. These fixed surfaces, or stabilizers, are incorporated into gyroplane designs to increase the pitch stability of the aircraft. Some gyroplanes use very little, if any, horizontal surface. This translates into less stability, but a higher degree of maneuverability. When used, a moveable horizontal surface, or elevator, adds additional pitch control of the aircraft. On early tractor configured gyroplanes, the elevator served an additional function of deflecting the propeller slipstream up and through the rotor to assist in prerotation.

COLLECTIVE CONTROL

The collective control provides a means to vary the rotor blade pitch of all the blades at the same time, and is available only on more advanced gyroplanes. When incorporated into the rotor head design, the collective allows jump takeoffs when the blade inertia is sufficient. Also, control of in-flight rotor r.p.m. is available to enhance cruise and landing performance. A simple two position collective does not allow unlimited control of blade pitch, but instead has one position for prerotation and another position for flight. This is a performance compromise but reduces pilot workload by simplifying control of the rotor system.

Gyroplane Systems

Gyroplanes are available in a wide variety of designs that range from amateur built to FAA-certificated aircraft. Similarly, the complexity of the systems integrated in gyroplane design cover a broad range. To ensure the airworthiness of your aircraft, it is important that you thoroughly understand the design and operation of each system employed by your machine.

PROPULSION SYSTEMS

Most of the gyroplanes flying today use a reciprocating engine mounted in a pusher configuration that drives either a fixed or constant speed propeller. The engines used in amateur-built gyroplanes are normally proven powerplants adapted from automotive or other uses. Some amateur-built gyroplanes use FAA-certificated aircraft engines and propellers. Auto engines, along with some of the other powerplants adapted to gyroplanes, operate at a high r.p.m., which requires the use of a reduction unit to lower the output to efficient propeller speeds.

Early autogyros used existing aircraft engines, which drove a propeller in the tractor configuration. Several amateur-built gyroplanes still use this propulsion configuration, and may utilize a certificated or an uncertificated engine. Although not in use today, turboprop and pure jet engines could also be used for the propulsion of a gyroplane.

ROTOR SYSTEMS
SEMIRIGID ROTOR SYSTEM

Any rotor system capable of autorotation may be utilized in a gyroplane. Because of its simplicity, the most widely used system is the semirigid, teeter-head system. This system is found in most amateur-built gyroplanes. [Figure 18-1] In this system, the rotor head is mounted on a spindle, which may be tilted for control. The rotor blades are attached to a hub bar that may or may not have adjustments for varying the blade pitch. A **coning angle**, determined by projections of blade weight, rotor speed, and load to be carried, is built into the hub bar. This minimizes hub bar bending moments and eliminates the need for a coning hinge, which is used in more complex rotor systems. A tower block provides the **undersling** and attachment to the rotor head by the teeter bolt. The rotor head is comprised of a bearing block in which the bearing is mounted and onto which the tower plates are attached. The spindle (commonly, a vertically oriented bolt) attaches the

rotating portion of the head to the non-rotating torque tube. The torque tube is mounted to the airframe through attachments allowing both lateral and longitudinal movement. This allows the movement through which control is achieved.

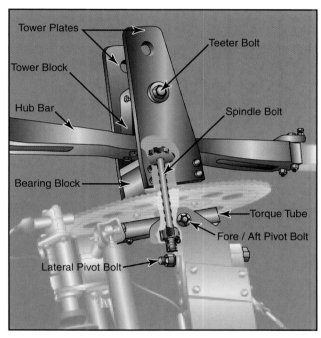

Figure 18-1. The semirigid, teeter-head system is found on most amateur-built gyroplanes. The rotor hub bar and blades are permitted to tilt by the teeter bolt.

FULLY ARTICULATED ROTOR SYSTEM

The fully articulated rotor system is found on some gyroplanes. As with helicopter-type rotor systems, the articulated rotor system allows the manipulation of

Coning Angle—An angular deflection of the rotor blades upward from the rotor hub.

Undersling—A design characteristic that prevents the distance between the rotor mast axis and the center of mass of each rotor blade from changing as the blades teeter. This precludes Coriolis Effect from acting on the speed of the rotor system. Undersling is further explained in Chapter 3—Aerodynamics of Flight, Coriolis Effect (Law of Conservation of Angular Momentum).

rotor blade pitch while in flight. This system is significantly more complicated than the teeter-head, as it requires hinges that allow each rotor blade to flap, feather, and lead or lag independently. [Figure 18-2] When used, the fully articulated rotor system of a gyroplane is very similar to those used on helicopters, which is explained in depth in Chapter 5—Helicopter Systems, Main Rotor Systems. One major advantage of using a fully articulated rotor in gyroplane design is that it usually allows jump takeoff capability. Rotor characteristics required for a successful jump takeoff must include a method of collective pitch change, a blade with sufficient inertia, and a prerotation mechanism capable of approximately 150 percent of rotor flight r.p.m.

Figure 18-2. The fully articulated rotor system enables the pilot to effect changes in pitch to the rotor blades, which is necessary for jump takeoff capability.

Incorporating rotor blades with high inertia potential is desirable in helicopter design and is essential for jump takeoff gyroplanes. A rotor hub design allowing the rotor speed to exceed normal flight r.p.m. by over 50 percent is not found in helicopters, and predicates a rotor head design particular to the jump takeoff gyroplane, yet very similar to that of the helicopter.

PREROTATOR

Prior to takeoff, the gyroplane rotor must first achieve a rotor speed sufficient to create the necessary lift. This is accomplished on very basic gyroplanes by initially spinning the blades by hand. The aircraft is then taxied with the rotor disc tilted aft, allowing airflow through the system to accelerate it to flight r.p.m. More advanced gyroplanes use a prerotator, which provides a mechanical means to spin the rotor. Many prerotators are capable of only achieving a portion of the speed necessary for flight; the remainder is gained by taxiing or during the takeoff roll. Because of the wide variety of prerotation systems available, you need to become thoroughly familiar with the characteristics and techniques associated with your particular system.

MECHANICAL PREROTATOR

Mechanical prerotators typically have clutches or belts for engagement, a drive train, and may use a transmission to transfer engine power to the rotor. Friction drives and flex cables are used in conjunction with an automotive type bendix and ring gear on many gyroplanes. [Figure 18-3]

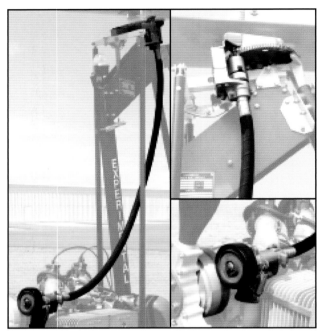

Figure 18-3. The mechanical prerotator used by many gyroplanes uses a friction drive at the propeller hub, and a flexible cable that runs from the propeller hub to the rotor mast. When engaged, the bendix spins the ring gear located on the rotor hub.

The mechanical prerotator used on jump takeoff gyroplanes may be regarded as being similar to the helicopter main rotor drive train, but only operates while the aircraft is firmly on the ground. Gyroplanes do not have an antitorque device like a helicopter, and ground contact is necessary to counteract the torque forces generated by the prerotation system. If jump takeoff capability is designed into a gyroplane, rotor r.p.m. prior to liftoff must be such that rotor energy will support the aircraft through the acceleration phase of takeoff. This combination of rotor system and prerotator utilizes the transmission only while the aircraft is on the ground, allowing the transmission to be disconnected from both the rotor and the engine while in normal flight.

HYDRAULIC PREROTATOR

The hydraulic prerotator found on gyroplanes uses engine power to drive a hydraulic pump, which in turn drives a hydraulic motor attached to an automotive type bendix and ring gear. [Figure 18-4] This system also requires that some type of clutch and pressure regulation be incorporated into the design.

Figure 18-4. This prerotator uses belts at the propeller hub to drive a hydraulic pump, which drives a hydraulic motor on the rotor mast.

ELECTRIC PREROTATOR

The electric prerotator found on gyroplanes uses an automotive type starter with a bendix and ring gear mounted at the rotor head to impart torque to the rotor system. [Figure 18-5] This system has the advantage of simplicity and ease of operation, but is dependent on having electrical power available. Using a "soft start" device can alleviate the problems associated with the high starting torque initially required to get the rotor system turning. This device delivers electrical pulses to the starter for approximately 10 seconds before connecting uninterrupted voltage.

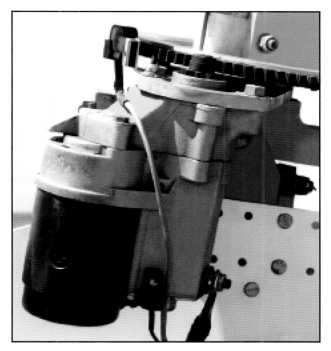

Figure 18-5. The electric prerotator is simple and easy to use, but requires the availability of electrical power.

TIP JETS

Jets located at the rotor blade tips have been used in several applications for prerotation, as well as for hover flight. This system has no requirement for a transmission or clutches. It also has the advantage of not imparting torque to the airframe, allowing the rotor to be powered in flight to give increased climb rates and even the ability to hover. The major disadvantage is the noise generated by the jets. Fortunately, tip jets may be shut down while operating in the autorotative gyroplane mode.

INSTRUMENTATION

The instrumentation required for flight is generally related to the complexity of the gyroplane. Some gyroplanes using air-cooled and fuel/oil-lubricated engines may have limited instrumentation.

ENGINE INSTRUMENTS

All but the most basic engines require monitoring instrumentation for safe operation. Coolant temperature, cylinder head temperatures, oil temperature, oil pressure, carburetor air temperature, and exhaust gas temperature are all direct indications of engine operation and may be displayed. Engine power is normally indicated by engine r.p.m., or by manifold pressure on gyroplanes with a constant speed propeller.

ROTOR TACHOMETER

Most gyroplanes are equipped with a rotor r.p.m. indicator. Because the pilot does not normally have direct control of rotor r.p.m. in flight, this instrument is most useful on the takeoff roll to determine when there is sufficient rotor speed for liftoff. On gyroplanes not equipped with a rotor tachometer, additional piloting skills are required to sense rotor r.p.m. prior to takeoff.

Certain gyroplane maneuvers require you to know precisely the speed of the rotor system. Performing a jump takeoff in a gyroplane with collective control is one example, as sufficient rotor energy must be available for the successful outcome of the maneuver. When variable collective and a rotor tachometer are used, more efficient rotor operation may be accomplished by using the lowest practical rotor r.p.m. [Figure 18-6]

Figure 18-6. A rotor tachometer can be very useful to determine when rotor r.p.m. is sufficient for takeoff.

SLIP/SKID INDICATOR

A yaw string attached to the nose of the aircraft and a conventional inclinometer are often used in gyroplanes to assist in maintaining coordinated flight. [Figure 18-7]

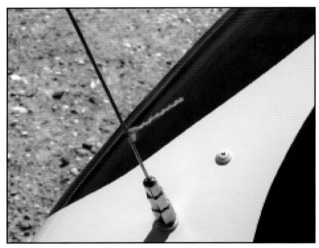

Figure 18-7. A string simply tied near the nose of the gyroplane that can be viewed from the cockpit is often used to indicate rotation about the yaw axis. An inclinometer may also be used.

AIRSPEED INDICATOR

Airspeed knowledge is essential and is most easily obtained by an airspeed indicator that is designed for accuracy at low airspeeds. Wind speed indicators have been adapted to many gyroplanes. When no air-speed indicator is used, as in some very basic amateur-built machines, you must have a very acute sense of "q" (impact air pressure against your body).

ALTIMETER

For the average pilot, it becomes increasingly difficult to judge altitude accurately when more than several hundred feet above the ground. A conventional altimeter may be used to provide an altitude reference when flying at higher altitudes where human perception degrades.

IFR FLIGHT INSTRUMENTATION

Gyroplane flight into instrument meteorological conditions requires adequate flight instrumentation and navigational systems, just as in any aircraft. Very few gyroplanes have been equipped for this type of operation. The majority of gyroplanes do not meet the stability requirements for single-pilot IFR flight. As larger and more advanced gyroplanes are developed, issues of IFR flight in these aircraft will have to be addressed.

GROUND HANDLING

The gyroplane is capable of ground taxiing in a manner similar to that of an airplane. A steerable nose wheel, which may be combined with independent main wheel brakes, provides the most common method of control. [Figure 18-8] The use of independent main wheel brakes allows differential braking, or applying more braking to one wheel than the other to achieve tight radius turns. On some gyroplanes, the steerable nose wheel is equipped with a foot-operated brake rather than using main wheel brakes. One limitation of this system is that the nose wheel normally supports only a fraction of the weight of the gyroplane, which greatly reduces braking effectiveness. Another drawback is the

Figure 18-8. Depending on design, main wheel brakes can be operated either independently or collectively. They are considerably more effective than nose wheel brakes.

inability to use differential braking, which increases the radius of turns.

The rotor blades demand special consideration during ground handling, as turning rotor blades can be a hazard to those nearby. Many gyroplanes have a rotor brake that may be used to slow the rotor after landing, or to secure the blades while parked. A parked gyroplane should never be left with unsecured blades, because even a slight change in wind could cause the blades to turn or flap.

CHAPTER 19

Gyroplane

Rotorcraft Flight Manual

As with most certificated aircraft manufactured after March 1979, FAA-certificated gyroplanes are required to have an approved flight manual. The flight manual describes procedures and limitations that must be adhered to when operating the aircraft. *Specification for Pilot's Operating Handbook*, published by the General Aviation Manufacturers Association (GAMA), provides a recommended format that more recent gyroplane flight manuals follow. [Figure 19-1]

ROTORCRAFT FLIGHT MANUAL

GENERAL—Presents basic information, such as loading, handling, and preflight of the gyroplane. Also includes definitions, abbreviations, symbology, and terminology explanations.

LIMITATIONS—Includes operating limitations, instrument markings, color coding, and basic placards necessary for the safe operation of the gyroplane.

EMERGENCY PROCEDURES—Provides checklists followed by amplified procedures for coping with various types of emergencies or critical situations. Related recommended airspeeds are also included. At the manufacturer's option, a section of abnormal procedures may be included to describe recommendations for handling equipment malfunctions or other abnormalities that are not of an emergency nature.

NORMAL PROCEDURES—Includes checklists followed by amplified procedures for conducting normal operations. Related recommended airspeeds are also provided.

PERFORMANCE—Gives performance information appropriate to the gyroplane, plus optional information presented in the most likely order for use in flight.

WEIGHT AND BALANCE—Includes weighing procedures, weight and balance records, computation instructions, and the equipment list.

AIRCRAFT AND SYSTEMS DESCRIPTION—Describes the gyroplane and its systems in a format considered by the manufacturer to be most informative.

HANDLING, SERVICE, AND MAINTENANCE—Includes information on gyroplane inspection periods, preventative maintenance that can be performed by the pilot, ground handling procedures, servicing, cleaning, and care instructions.

SUPPLEMENTS—Contains information necessary to safely and efficiently operate the gyroplane's various optional systems and equipment.

SAFETY AND OPERATIONAL TIPS—Includes optional information from the manufacturer of a general nature addressing safety practices and procedures.

Figure 19-1. The FAA-approved flight manual may contain as many as ten sections, as well as an optional alphabetical index.

This format is the same as that used by helicopters, which is explained in depth in Chapter 6—Rotorcraft Flight Manual (Helicopter).

Amateur-built gyroplanes may have operating limitations but are not normally required to have an approved flight manual. One exception is an exemption granted by the FAA that allows the commercial use of two-place, amateur-built gyroplanes for instructional purposes. One of the conditions of this exemption is to have an approved flight manual for the aircraft. This manual is to be used for training purposes, and must be carried in the gyroplane at all times.

USING THE FLIGHT MANUAL
The flight manual is required to be on board the aircraft to guarantee that the information contained therein is readily available. For the information to be of value, you must be thoroughly familiar with the manual and be able to read and properly interpret the various charts and tables.

WEIGHT AND BALANCE SECTION
The weight and balance section of the flight manual contains information essential to the safe operation of the gyroplane. Careful consideration must be given to the weight of the passengers, baggage, and fuel prior to each flight. In conducting weight and balance computations, many of the terms and procedures are similar to those used in helicopters. These are further explained in Chapter 7—Weight and Balance. In any aircraft, failure to adhere to the weight and balance limitations prescribed by the manufacturer can be extremely hazardous.

SAMPLE PROBLEM
As an example of a weight and balance computation, assume a sightseeing flight in a two-seat, tandem-configured gyroplane with two people aboard. The pilot, seated in the front, weighs 175 pounds while the rear seat passenger weighs 160 pounds. For the purposes of this example, there will be no baggage carried. The basic empty weight of the aircraft is 1,315 pounds with a moment, divided by 1,000, of 153.9 pound-inches.

Using the loading graph [Figure 19-2], the moment/1000 of the pilot is found to be 9.1 pound-inches, and the passenger has a moment/1000 of 13.4 pound-inches.

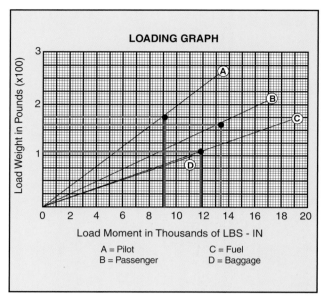

Figure 19-2. A loading graph is used to determine the load moment for weights at various stations.

Adding these figures, the total weight of the aircraft for this flight (without fuel) is determined to be 1,650 pounds with a moment/1000 of 176.4 pound-inches. [Figure 19-3]

	Weight (pounds)	Moment (lb.-in./1,000)
Basic Empty Weight	1,315	153.9
Pilot	175	9.1
Passenger	160	13.4
Baggage	0	0
Total Aircraft (Less Fuel)	**1,650**	**176.4**
Max Gross Weight = **1,800 lbs**.		

Figure 19-3. Loading of the sample aircraft, less fuel.

The maximum gross weight for the sample aircraft is 1,800 pounds, which allows up to 150 pounds to be carried in fuel. For this flight, 18 gallons of fuel is deemed sufficient. Allowing six pounds per gallon of fuel, the fuel weight on the aircraft totals 108 pounds. Referring again to the loading graph [Figure 19-2], 108 pounds of fuel would have a moment/1000 of 11.9 pound-inches. This is added to the previous totals to obtain the total aircraft weight of 1,758 pounds and a moment/1000 of 188.3. Locating this point on the center of gravity envelope chart [Figure 19-4], shows that the loading is within the prescribed weight and balance limits.

	Weight (lbs.)	Moment (lb.-ins. /1,000)
1. Total Aircraft Weight (Less Fuel)	1,650	176.4
3. Fuel	108	11.9
TOTALS	1,758	188.3

CENTER OF GRAVITY ENVELOPE

Figure 19-4. Center of gravity envelope chart.

PERFORMANCE SECTION

The performance section of the flight manual contains data derived from actual flight testing of the aircraft. Because the actual performance may differ, it is prudent to maintain a margin of safety when planning operations using this data.

SAMPLE PROBLEM

For this example, a gyroplane at its maximum gross weight (1,800 lbs.) needs to perform a short field take-off due to obstructions in the takeoff path. Present weather conditions are standard temperature at a pressure altitude of 2,000 feet, and the wind is calm. Referring to the appropriate performance chart [Figure 19-5], the takeoff distance to clear a 50-foot obstacle is determined by entering the chart from the left at the pressure altitude of 2,000 feet. You then proceed horizontally to the right until intersecting the appropriate temperature reference line, which in this case is the dashed standard temperature line. From this point, descend vertically to find the total takeoff distance to clear a 50-foot obstacle. For the conditions given, this particular gyroplane would require a distance of 940 feet for ground roll and the distance needed to climb 50 feet above the surface. Notice that the data presented in this chart is predicated on certain conditions, such as a running takeoff to 30 m.p.h., a 50 m.p.h. climb speed, a

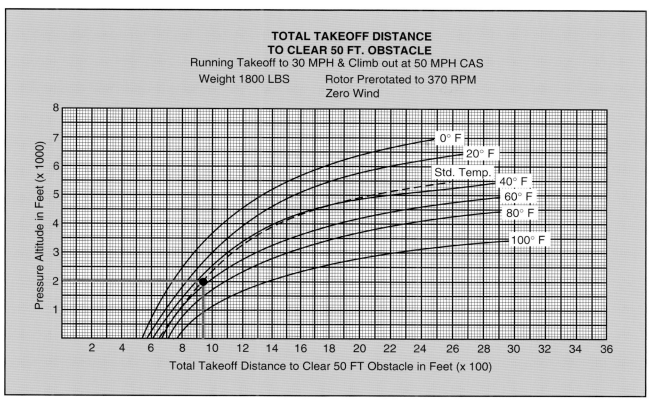

Figure 19-5. Takeoff performance chart.

rotor prerotation speed of 370 r.p.m., and no wind. Variations from these conditions alter performance, possibly to the point of jeopardizing the successful outcome of the maneuver.

HEIGHT/VELOCITY DIAGRAM

Like helicopters, gyroplanes have a height/velocity diagram that defines what speed and altitude combinations allow for a safe landing in the event of an engine failure. [Figure 19-6]

During an engine-out landing, the cyclic flare is used to arrest the vertical velocity of the aircraft and most of the forward velocity. On gyroplanes with a manual collective control, increasing blade pitch just prior to touchdown can further reduce ground roll. Typically, a gyroplane has a lower rotor disc loading than a helicopter, which provides a slower rate of descent in autorotation. The power required to turn the main transmission, tail rotor transmission, and tail rotor also add to the higher descent rate of a helicopter in autorotation as compared with that of a gyroplane.

EMERGENCY SECTION

Because in-flight emergencies may not allow enough time to reference the flight manual, the emergency section should be reviewed periodically to maintain familiarity with these procedures. Many aircraft also use placards and instrument markings in the cockpit, which provide important information that may not be committed to memory.

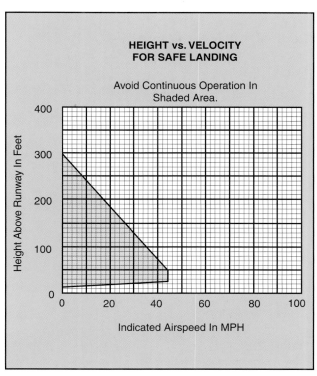

Figure 19-6. Operations within the shaded area of a height/velocity diagram may not allow for a safe landing and are to be avoided.

Hang Test

The proper weight and balance of a gyroplane without a flight manual is normally determined by conducting a hang test of the aircraft. This is achieved by removing the rotor blades and suspending the aircraft by its teeter bolt, free from contact with the ground. A measurement is then taken, either at the keel or the rotor mast, to determine how many degrees from level the gyroplane hangs. This number must be within the range specified by the manufacturer. For the test to reflect the true balance of the aircraft, it is important that it be conducted using the actual weight of the pilot and all gear normally carried in flight. Additionally, the measurement should be taken both with the fuel tank full and with it empty to ensure that fuel burn does not affect the loading.

The diversity of gyroplane designs available today yields a wide variety of capability and performance. For safe operation, you must be thoroughly familiar with the procedures and limitations for your particular aircraft along with other factors that may affect the safety of your flight.

PREFLIGHT

As pilot in command, you are the final authority in determining the airworthiness of your aircraft. Adherence to a preflight checklist greatly enhances your ability to evaluate the fitness of your gyroplane by ensuring that a complete and methodical inspection of all components is performed. [Figure 20-1] For aircraft without a formal checklist, it is prudent to create one that is specific to the aircraft to be sure that important items are not overlooked. To determine the status of required inspections, a preflight review of the aircraft records is also necessary.

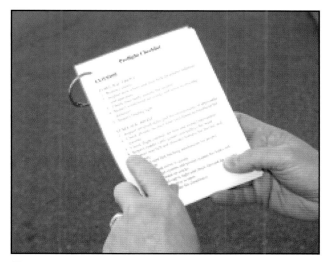

Figure 20-1. A checklist is extremely useful in conducting a thorough preflight inspection.

COCKPIT MANAGEMENT

As in larger aircraft, cockpit management is an important skill necessary for the safe operation of a gyroplane. Intrinsic to these typically small aircraft is a limited amount of space that must be utilized to its potential. The placement and accessibility of charts, writing materials, and other necessary items must be carefully considered. Gyroplanes with open cockpits add the challenge of coping with wind, which further increases the need for creative and resourceful cockpit management for optimum efficiency.

ENGINE STARTING

The dissimilarity between the various types of engines used for gyroplane propulsion necessitates the use of an engine start checklist. Again, when a checklist is not provided, it is advisable to create one for the safety of yourself and others, and to prevent inadvertent damage to the engine or propeller. Being inherently dangerous, the propeller demands special attention during engine starting procedures. Always ensure that the propeller area is clear prior to starting. In addition to providing an added degree of safety, being thoroughly familiar with engine starting procedures and characteristics can also be very helpful in starting an engine under various weather conditions.

TAXIING

The ability of the gyroplane to be taxied greatly enhances its utility. However, a gyroplane should not be taxied in close proximity to people or obstructions while the rotor is turning. In addition, taxi speed should be limited to no faster than a brisk walk in ideal conditions, and adjusted appropriately according to the circumstances.

BLADE FLAP

On a gyroplane with a semi-rigid, teeter-head rotor system, blade flap may develop if too much airflow passes through the rotor system while it is operating at low r.p.m. This is most often the result of taxiing too fast for a given rotor speed. Unequal lift acting on the advancing and retreating blades can cause the blades to teeter to the maximum allowed by the rotor head design. The blades then hit the teeter stops, creating a vibration that may be felt in the cyclic control. The frequency of the vibration corresponds to the speed of the rotor, with the blades hitting the stops twice during each revolution. If the flapping is not controlled, the situation can grow worse as the blades begin to flex and bend. Because the system is operating at low r.p.m., there is not enough centrifugal force acting on the blades to keep them rigid. The shock of hitting the teeter stops combined with uneven lift along the length of the blade causes an undulation to begin, which can increase in severity if allowed to progress. In extreme cases, a rotor blade may strike the ground or propeller. [Figure 20-2]

Figure 20-2. Taxiing too fast or gusting winds can cause blade flap in a slow turning rotor. If not controlled, a rotor blade may strike the ground.

To avoid the onset of blade flap, always taxi the gyroplane at slow speeds when the rotor system is at low r.p.m. Consideration must also be given to wind speed and direction. If taxiing into a 10-knot headwind, for example, the airflow through the rotor will be 10 knots faster than the forward speed of the gyroplane, so the taxi speed should be adjusted accordingly. When prerotating the rotor by taxiing with the rotor disc tilted aft, allow the rotor to accelerate slowly and smoothly. In the event blade flap is encountered, apply forward cyclic to reduce the rotor disc angle and slow the gyroplane by reducing throttle and applying the brakes, if needed. [Figure 20-3]

BEFORE TAKEOFF

For the amateur-built gyroplane using single ignition and a fixed trim system, the before takeoff check is quite simple. The engine should be at normal operating temperature, and the area must be clear for prerotation. Certificated gyroplanes using conventional aircraft engines have a checklist that includes items specific to the powerplant. These normally include, but are not limited to, checks for magneto drop, carburetor heat, and, if a constant speed propeller is installed, that it be cycled for proper operation.

Following the engine run-up is the procedure for accomplishing prerotation. This should be reviewed and committed to memory, as it typically requires both hands to perform.

PREROTATION

Prerotation of the rotor can take many forms in a gyroplane. The most basic method is to turn the rotor blades by hand. On a typical gyroplane with a counter-clockwise rotating rotor, prerotation by hand is done on the right side of the rotor disk. This allows body movement to be directed away from the propeller to minimize the risk of injury. Other methods of prerotation include using mechanical, electrical, or hydraulic means for the initial blade spin-up. Many of these systems can achieve only a portion of the rotor speed that is necessary for takeoff. After the prerotator is disengaged, taxi the gyroplane with the rotor disk tilted aft to allow airflow through the rotor. This increases rotor speed to flight r.p.m. In windy conditions, facing the gyroplane into the wind during prerotation assists in achieving the highest possible rotor speed from the prerotator. A factor often overlooked that can negatively affect the prerotation speed is the cleanliness of the rotor blades. For maximum efficiency, it is recommended that the rotor blades be cleaned periodically. By obtaining the maximum possible rotor speed through the use of proper prerotation techniques, you

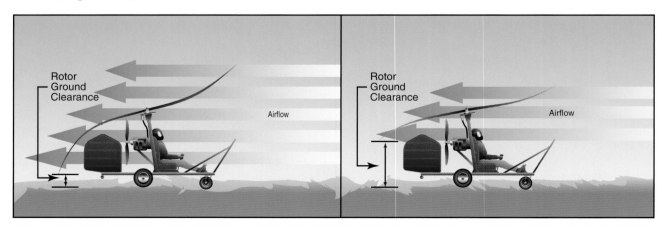

Figure 20-3. Decreasing the rotor disc angle of attack with forward cyclic can reduce the excessive amount of airflow causing the blade flap. This also allows greater clearance between the rotor blades and the surface behind the gyroplane, minimizing the chances of a blade striking the ground.

minimize the length of the ground roll that is required to get the gyroplane airborne.

The prerotators on certificated gyroplanes remove the possibility of blade flap during prerotation. Before the clutch can be engaged, the pitch must be removed from the blades. The rotor is then prerotated with a 0° angle of attack on the blades, which prevents lift from being produced and precludes the possibility of flapping. When the desired rotor speed is achieved, blade pitch is increased for takeoff.

TAKEOFF

Takeoffs are classified according to the takeoff surface, obstructions, and atmospheric conditions. Each type of takeoff assumes that certain conditions exist. When conditions dictate, a combination of takeoff techniques can be used. Two important speeds used for takeoff and initial climbout are V_X and V_Y. V_X is defined as the speed that provides the best angle of climb, and will yield the maximum altitude gain over a given distance. This speed is normally used when obstacles on the ground are a factor. Maintaining V_Y speed ensures the aircraft will climb at its maximum rate, providing the most altitude gain for a given period of time. [Figure 20-4] Prior to any takeoff or maneuver, you should ensure that the area is clear of other traffic.

NORMAL TAKEOFF

The normal takeoff assumes that a prepared surface of adequate length is available and that there are no high obstructions to be cleared within the takeoff path. The normal takeoff for most amateur-built gyroplanes is accomplished by prerotating to sufficient rotor r.p.m. to prevent blade flapping and tilting the rotor back with cyclic control. Using a speed of 20 to 30 m.p.h., allow the rotor to accelerate and begin producing lift. As lift increases, move the cyclic forward to decrease the pitch angle on the rotor disc. When appreciable lift is being produced, the nose of the aircraft rises, and you can feel an increase in drag. Using coordinated throttle and flight control inputs, balance the gyroplane on the main gear without the nose wheel or tail wheel in contact with the surface. At this point, smoothly increase power to full thrust and hold the nose at takeoff attitude with cyclic pressure. The gyroplane will lift off at or near the minimum power required speed for the aircraft. V_X should be used for the initial climb, then V_Y for the remainder of the climb phase.

A normal takeoff for certificated gyroplanes is accomplished by prerotating to a rotor r.p.m. slightly above that required for flight and disengaging the rotor drive. The brakes are then released and full power is applied. Lift off will not occur until the blade pitch is increased to the normal in-flight setting and the rotor disk tilted

Figure 20-4. Best angle-of-climb (V_X) speed is used when obstacles are a factor. V_Y provides the most altitude gain for a given amount of time.

aft. This is normally accomplished at approximately 30 to 40 m.p.h. The gyroplane should then be allowed to accelerate to V_X for the initial climb, followed by V_Y for the remainder of the climb. On any takeoff in a gyroplane, engine torque causes the aircraft to roll opposite the direction of propeller rotation, and adequate compensation must be made.

CROSSWIND TAKEOFF

A crosswind takeoff is much like a normal takeoff, except that you have to use the flight controls to compensate for the crosswind component. The term crosswind component refers to that part of the wind which acts at right angles to the takeoff path. Before attempting any crosswind takeoff, refer to the flight manual, if available, or the manufacturer's recommendations for any limitations.

Begin the maneuver by aligning the gyroplane into the wind as much as possible. At airports with wide runways, you might be able to angle your takeoff roll down the runway to take advantage of as much headwind as you can. As airspeed increases, gradually tilt the rotor into the wind and use rudder pressure to maintain runway heading. In most cases, you should accelerate to a speed slightly faster than normal liftoff speed. As you reach takeoff speed, the downwind wheel lifts off the ground first, followed by the upwind wheel. Once airborne, remove the cross-control inputs and establish a crab, if runway heading is to be maintained. Due to the maneuverability of the gyroplane, an immediate turn into the wind after lift off can be safely executed, if this does not cause a conflict with existing traffic.

COMMON ERRORS FOR NORMAL AND CROSSWIND TAKEOFFS

1. Failure to check rotor for proper operation, track, and r.p.m. prior to takeoff.

2. Improper initial positioning of flight controls.

3. Improper application of power.

4. Poor directional control.

5. Failure to lift off at proper airspeed.

6. Failure to establish and maintain proper climb attitude and airspeed.

7. Drifting from the desired ground track during the climb.

SHORT-FIELD TAKEOFF

Short-field takeoff and climb procedures may be required when the usable takeoff surface is short, or when it is restricted by obstructions, such as trees, powerlines, or buildings, at the departure end. The technique is identical to the normal takeoff, with performance being optimized during each phase. Using the help from wind and propwash, the maximum rotor r.p.m. should be attained from the prerotator and full power applied as soon as appreciable lift is felt. V_X climb speed should be maintained until the obstruction is cleared. Familiarity with the rotor acceleration characteristics and proper technique are essential for optimum short-field performance.

If the prerotator is capable of spinning the rotor in excess of normal flight r.p.m., the stored energy may be used to enhance short-field performance. Once maximum rotor r.p.m. is attained, disengage the rotor drive, release the brakes, and apply power. As airspeed and rotor r.p.m. increase, apply additional power until full power is achieved. While remaining on the ground, accelerate the gyroplane to a speed just prior to V_X. At that point, tilt the disk aft and increase the blade pitch to the normal in-flight setting. The climb should be at a speed just under V_X until rotor r.p.m. has dropped to normal flight r.p.m. or the obstruction has been cleared. When the obstruction is no longer a factor, increase the airspeed to V_Y.

COMMON ERRORS

1. Failure to position gyroplane for maximum utilization of available takeoff area.

2. Failure to check rotor for proper operation, track, and r.p.m. prior to takeoff.

3. Improper initial positioning of flight controls.

4. Improper application of power.

5. Improper use of brakes.

6. Poor directional control.

7. Failure to lift off at proper airspeed.

8. Failure to establish and maintain proper climb attitude and airspeed.

9. Drifting from the desired ground track during the climb.

HIGH-ALTITUDE TAKEOFF

A high-altitude takeoff is conducted in a manner very similar to that of the short-field takeoff, which achieves maximum performance from the aircraft during each phase of the maneuver. One important consideration is that at higher altitudes, rotor r.p.m. is higher for a given blade pitch angle. This higher speed is a result of thinner air, and is necessary to produce the same amount of lift. The inertia of the excess rotor speed should not be used in an attempt to enhance climb performance. Another important consideration is the effect of altitude on engine performance. As altitude increases, the amount of oxygen available for combustion decreases. In **normally aspirated** engines, it may be necessary to

---

Normally Aspirated—An engine that does not compensate for decreases in atmospheric pressure through turbocharging or other means.

adjust the fuel/air mixture to achieve the best possible power output. This process is referred to as "leaning the mixture." If you are considering a high-altitude takeoff, and it appears that the climb performance limit of the gyroplane is being approached, do not attempt a takeoff until more favorable conditions exist.

SOFT-FIELD TAKEOFF

A soft field may be defined as any takeoff surface that measurably retards acceleration during the takeoff roll. The objective of the soft-field takeoff is to transfer the weight of the aircraft from the landing gear to the rotor as quickly and smoothly as possible to eliminate the drag caused by surfaces, such as tall grass, soft dirt, or snow. This takeoff requires liftoff at a speed just above the minimum level flight speed for the aircraft. Due to design, many of the smaller gyroplanes have a limited pitch attitude available, as tail contact with the ground prevents high pitch attitudes until in flight. At minimum level flight speed, the pitch attitude is often such that the tail wheel is lower than the main wheels. When performing a soft-field takeoff, these aircraft require slightly higher liftoff airspeeds to allow for proper tail clearance.

COMMON ERRORS

1. Failure to check rotor for proper operation, track, and r.p.m. prior to takeoff.

2. Improper initial positioning of flight controls.

3. Improper application of power.

4. Allowing gyroplane to lose momentum by slowing or stopping on takeoff surface prior to initiating takeoff.

5. Poor directional control.

6. Improper pitch attitude during lift-off.

7. Settling back to takeoff surface after becoming airborne.

8. Failure to establish and maintain proper climb attitude and airspeed.

9. Drifting from the desired ground track during the climb.

JUMP TAKEOFF

Gyroplanes with collective pitch change, and the ability to prerotate the rotor system to speeds approximately 50 percent higher than those required for normal flight, are capable of achieving extremely short takeoff rolls. Actual jump takeoffs can be performed under the proper conditions. A jump takeoff requires no ground roll, making it the most effective soft-field and crosswind takeoff procedure. [Figure 20-5] A jump takeoff is possible because the energy stored in the blades, as a result of the higher rotor r.p.m., is used to keep the gyroplane airborne as it accelerates through

Figure 20-5. During a jump takeoff, excess rotor inertia is used to lift the gyroplane nearly vertical, where it is then accelerated through minimum level flight speed.

minimum level flight speed. Failure to have sufficient rotor r.p.m. for a jump takeoff results in the gyroplane settling back to the ground. Before attempting a jump takeoff, it is essential that you first determine if it is possible given the existing conditions by consulting the relevant performance chart. Should conditions of weight, altitude, temperature, or wind leave the successful outcome of the maneuver in doubt, it should not be attempted.

The prudent pilot may also use a "rule of thumb" for predicting performance before attempting a jump takeoff. As an example, suppose that a particular gyroplane is known to be able to make a jump takeoff and remain airborne to accelerate to V_X at a weight of 1,800 pounds and a **density altitude** of 2,000 feet. Since few takeoffs are made under these exact conditions, compensation must be made for variations in weight, wind, and density altitude. The "rule of thumb" being used for this particular aircraft stipulates that 1,000 feet of density altitude equates with 10 m.p.h. wind or 100 pounds of gross weight. To use this equation, you must first determine the density altitude. This is accomplished by setting your altimeter to the standard sea level pressure setting of 29.92 inches of mercury and reading the pressure altitude. Next, you must correct for nonstandard temperature. Standard temperature at sea level is 59°F (15°C) and decreases 3.5°F (2°C) for every additional

Density Altitude—Pressure altitude corrected for nonstandard temperature. This is a theoretical value that is used in determining aircraft performance.

one thousand feet of pressure altitude. [Figure 20-6] Once you have determined the standard temperature for your pressure altitude, compare it with the actual existing conditions. For every 10°F (5.5°C) the actual temperature is above standard, add 750 feet to the pressure altitude to estimate the density altitude. If the density altitude is above 2,000 feet, a jump takeoff in this aircraft should not be attempted unless wind and/or a weight reduction would compensate for the decrease in performance. Using the equation, if the density altitude is 3,000 feet (1,000 feet above a satisfactory jump density altitude), a reduction of 100 pounds in gross weight or a 10 m.p.h. of wind would still allow a satisfactory jump takeoff. Additionally, a reduction of 50 pounds in weight combined with a 5 m.p.h. wind would also allow a satisfactory jump. If it is determined that a jump takeoff should not be conducted because the weight cannot be reduced or an appropriate wind is not blowing, then consideration should be given to a rolling takeoff. A takeoff roll of 10 m.p.h. is equivalent to a wind speed of 10 m.p.h. or a reduction of 100 pounds in gross weight. It is important to note that a jump takeoff is predicated on having achieved a specific rotor r.p.m. If this r.p.m. has not been attained, performance is unpredictable, and the maneuver should not be attempted.

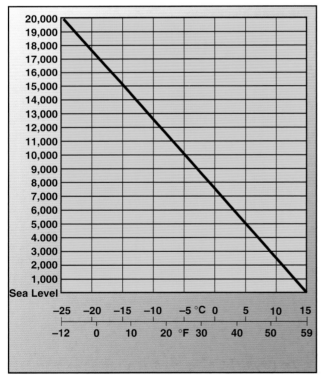

Figure 20-6. Standard temperature chart.

BASIC FLIGHT MANEUVERS
Conducting flight maneuvers in a gyroplane is different than in most other aircraft. Because of the wide variety in designs, many gyroplanes have only basic instruments available, and the pilot is often exposed to the airflow. In addition, the visual clues found on other aircraft, such as cowlings, wings, and windshields might not be part of your gyroplane's design. Therefore, much more reliance is placed on pilot interpretation of flight attitude and the "feel" of the gyroplane than in other types of aircraft. Acquiring the skills to precisely control a gyroplane can be a challenging and rewarding experience, but requires dedication and the direction of a competent instructor.

STRAIGHT-AND-LEVEL FLIGHT
Straight-and-level flight is conducted by maintaining a constant altitude and a constant heading. In flight, a gyroplane essentially acts as a plumb suspended from the rotor. As such, torque forces from the engine cause the airframe to be deflected a few degrees out of the vertical plane. This very slight "out of vertical" condition should be ignored and the aircraft flown to maintain a constant heading.

The throttle is used to control airspeed. In level flight, when the airspeed of a gyroplane increases, the rotor disc angle of attack must be decreased. This causes pitch control to become increasingly more sensitive. [Figure 20-7] As this disc angle becomes very small, it is possible to overcontrol a gyroplane when encountering turbulence. For this reason, when extreme turbulence is encountered or expected, airspeed should be decreased. Even in normal conditions, a gyroplane requires constant attention to maintain straight-and-level flight. Although more stable than helicopters, gyroplanes are less stable than airplanes. When cyclic trim is available, it should be used to relieve any stick forces required during stabilized flight.

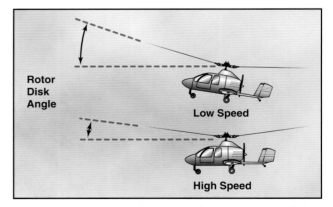

Figure 20-7. The angle of the rotor disc decreases at higher cruise speeds, which increases pitch control sensitivity.

CLIMBS
A climb is achieved by adding power in excess of what is required for straight-and-level flight at a particular airspeed. The amount of excess power used is directly proportional to the climb rate. For maneuvers when

maximum performance is desired, two important climb speeds are best angle-of-climb speed and best rate-of-climb speed.

Because a gyroplane cannot be stalled, it may be tempting to increase the climb rate by decreasing airspeed. This practice, however, is self-defeating. Operating below the best angle-of-climb speed causes a diminishing rate of climb. In fact, if a gyroplane is slowed to the minimum level flight speed, it requires full power just to maintain altitude. Operating in this performance realm, sometimes referred to as the "backside of the power curve," is desirable in some maneuvers, but can be hazardous when maximum climb performance is required. For further explanation of a gyroplane power curve, see Flight at Slow Airspeeds, which is discussed later in this chapter.

DESCENTS
A descent is the result of using less power than that required for straight-and-level flight at a particular airspeed. Varying engine power during a descent allows you to choose a variety of descent profiles. In a power-off descent, the minimum descent rate is achieved by using the airspeed that would normally be used for level flight at minimum power, which is also very close to the speed used for the best angle of climb. When distance is a factor during a power-off descent, maximum gliding distance can be achieved by maintaining a speed very close to the best rate-of-climb airspeed. Because a gyroplane can be safely flown down to zero airspeed, a common error in this type of descent is attempting to extend the glide by raising the pitch attitude. The result is a higher rate of descent and less distance being covered. For this reason, proper glide speed should be adhered to closely. Should a strong headwind exist, while attempting to achieve the maximum distance during a glide, a rule of thumb to achieve the greatest distance is to increase the glide speed by approximately 25 percent of the headwind. The attitude of the gyroplane for best glide performance is learned with experience, and slight pitch adjustments are made for the proper airspeed. If a descent is needed to lose excess altitude, slowing the gyroplane to below the best glide speed increases the rate of descent. Typically, slowing to zero airspeed results in a descent rate twice that of maintaining the best glide speed.

TURNS
Turns are made in a gyroplane by banking the rotor disc with cyclic control. Once the area, in the direction of the turn, has been cleared for traffic, apply sideward pressure on the cyclic until the desired bank angle is achieved. The speed at which the gyroplane enters the bank is dependent on how far the cyclic is displaced. When the desired bank angle is reached, return the cyclic to the neutral position. The rudder pedals are used to keep the gyroplane in longitudinal trim throughout the turn, but not to assist in establishing the turn.

The bank angle used for a turn directly affects the rate of turn. As the bank is steepened, the turn rate increases, but more power is required to maintain altitude. A bank angle can be reached where all available power is required, with any further increase in bank resulting in a loss of airspeed or altitude. Turns during a climb should be made at the minimum angle of bank necessary, as higher bank angles would require more power that would otherwise be available for the climb. Turns while gliding increase the rate of descent and may be used as an effective way of losing excess altitude.

SLIPS
A slip occurs when the gyroplane slides sideways toward the center of the turn. [Figure 20-8] It is caused by an insufficient amount of rudder pedal in the direction of the turn, or too much in the direction opposite the turn. In other words, holding improper rudder pedal pressure keeps the nose from following the turn, the gyroplane slips sideways toward the center of the turn.

Figure 20-8. During a slip, the rate of turn is too slow for the angle of bank used, and the horizontal component of lift (HCL) exceeds inertia. You can reestablish equilibrium by decreasing the angle of bank, increasing the rate of turn by applying rudder pedal, or a combination of the two.

SKIDS
A skid occurs when the gyroplane slides sideways away from the center of the turn. [Figure 20-9] It is caused by too much rudder pedal pressure in the direction of the turn, or by too little in the direction opposite the turn. If the gyroplane is forced to turn faster with increased pedal pressure instead of by increasing the degree of

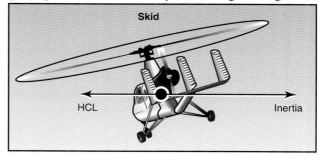

Figure 20-9. During a skid, inertia exceeds the HCL. To reestablish equilibrium, increase the bank angle or reduce the rate of turn by applying rudder pedal. You may also use a combination of these two corrections.

bank, it skids sideways away from the center of the turn instead of flying in its normal curved pattern.

COMMON ERRORS DURING BASIC FLIGHT MANEUVERS

1. Improper coordination of flight controls.

2. Failure to cross-check and correctly interpret outside and instrument references.

3. Using faulty trim technique.

STEEP TURNS

A steep turn is a performance maneuver used in training that consists of a turn in either direction at a bank angle of approximately 40°. The objective of performing steep turns is to develop smoothness, coordination, orientation, division of attention, and control techniques.

Prior to initiating a steep turn, or any other flight maneuver, first complete a clearing turn to check the area for traffic. To accomplish this, you may execute either one 180° turn or two 90° turns in opposite directions. Once the area has been cleared, roll the gyroplane into a 40° angle-of-bank turn while smoothly adding power and slowly moving the cyclic aft to maintain altitude. Maintain coordinated flight with proper rudder pedal pressure. Throughout the turn, cross-reference visual cues outside the gyroplane with the flight instruments, if available, to maintain a constant altitude and angle of bank. Anticipate the roll-out by leading the roll-out heading by approximately 20°. Using section lines or prominent landmarks to aid in orientation can be helpful in rolling out on the proper heading. During roll-out, gradually return the cyclic to the original position and reduce power to maintain altitude and airspeed.

COMMON ERRORS

1. Improper bank and power coordination during entry and rollout.

2. Uncoordinated use of flight controls.

3. Exceeding manufacturer's recommended maximum bank angle.

4. Improper technique in correcting altitude deviations.

5. Loss of orientation.

6. Excessive deviation from desired heading during rollout.

GROUND REFERENCE MANEUVERS

Ground reference maneuvers are training exercises flown to help you develop a division of attention between the flight path and ground references, while controlling the gyroplane and watching for other aircraft in the vicinity. Prior to each maneuver, a clearing turn should be accomplished to ensure the practice area is free of conflicting traffic.

RECTANGULAR COURSE

The rectangular course is a training maneuver in which the ground track of the gyroplane is equidistant from all sides of a selected rectangular area on the ground. [Figure 20-10] While performing the maneuver, the altitude and airspeed should be held constant. The rectangular course helps you to develop a recognition of a drift toward or away from a line parallel to the intended ground track. This is helpful in recognizing drift toward or from an airport runway during the various legs of the airport traffic pattern.

For this maneuver, pick a square or rectangular field, or an area bounded on four sides by section lines or roads, where the sides are approximately a mile in length. The area selected should be well away from other air traffic. Fly the maneuver approximately 600 to 1,000 feet above the ground, which is the altitude usually required for an airport traffic pattern. You should fly the gyroplane parallel to and at a uniform distance, about one-fourth to one-half mile, from the field boundaries, not above the boundaries. For best results, position your flight path outside the field boundaries just far enough away that they may be easily observed. You should be able to see the edges of the selected field while seated in a normal position and looking out the side of the gyroplane during either a left-hand or right-hand course. The distance of the ground track from the edges of the field should be the same regardless of whether the course is flown to the left or right. All turns should be started when your gyroplane is abeam the corners of the field boundaries. The bank normally should not exceed 30°.

Although the rectangular course may be entered from any direction, this discussion assumes entry on a downwind heading. As you approach the field boundary on the downwind leg, you should begin planning for your turn to the crosswind leg. Since you have a tailwind on the downwind leg, the gyroplane's groundspeed is increased (position 1). During the turn onto the crosswind leg, which is the equivalent of the base leg in a traffic pattern, the wind causes the gyroplane to drift away from the field. To counteract this effect, the roll-in should be made at a fairly fast rate with a relatively steep bank (position 2).

As the turn progresses, the tailwind component decreases, which decreases the groundspeed. Consequently, the bank angle and rate of turn must be reduced gradually to ensure that upon completion of the turn, the crosswind ground track continues to be the same distance from the edge of the field. Upon completion of the turn, the gyroplane should be level and

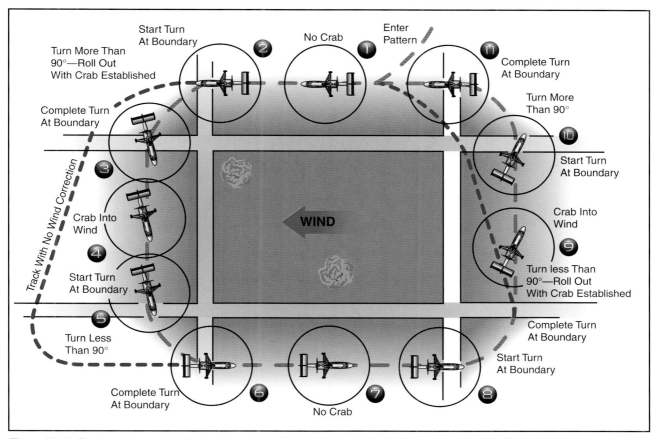

Figure 20-10. Rectangular course. The numbered positions in the text refer to the numbers in this illustration.

aligned with the downwind corner of the field. However, since the crosswind is now pushing you away from the field, you must establish the proper drift correction by flying slightly into the wind. Therefore, the turn to crosswind should be greater than a 90° change in heading (position 3). If the turn has been made properly, the field boundary again appears to be one-fourth to one-half mile away. While on the cross-wind leg, the wind correction should be adjusted, as necessary, to maintain a uniform distance from the field boundary (position 4).

As the next field boundary is being approached (position 5), plan the turn onto the upwind leg. Since a wind correction angle is being held into the wind and toward the field while on the crosswind leg, this next turn requires a turn of less than 90°. Since the crosswind becomes a headwind, causing the groundspeed to decrease during this turn, the bank initially must be medium and progressively decreased as the turn proceeds. To complete the turn, time the rollout so that the gyroplane becomes level at a point aligned with the corner of the field just as the longitudinal axis of the gyroplane again becomes parallel to the field boundary (position 6). The distance from the field boundary should be the same as on the other sides of the field.

On the upwind leg, the wind is a headwind, which results in an decreased groundspeed (position 7). Consequently, enter the turn onto the next leg with a fairly slow rate of roll-in, and a relatively shallow bank (position 8). As the turn progresses, gradually increase the bank angle because the headwind component is diminishing, resulting in an increasing groundspeed. During and after the turn onto this leg, the wind tends to drift the gyroplane toward the field boundary. To compensate for the drift, the amount of turn must be less than 90° (position 9).

Again, the rollout from this turn must be such that as the gyroplane becomes level, the nose of the gyroplane is turned slightly away the field and into the wind to correct for drift. The gyroplane should again be the same distance from the field boundary and at the same altitude, as on other legs. Continue the crosswind leg until the downwind leg boundary is approached (position 10). Once more you should anticipate drift and turning radius. Since drift correction was held on the crosswind leg, it is necessary to turn greater than 90° to align the gyroplane parallel to the downwind leg boundary. Start this turn with a medium bank angle, gradually increasing it to a steeper bank as the turn progresses. Time the rollout to assure paralleling the boundary of the field as the gyroplane becomes level (position 11).

If you have a direct headwind or tailwind on the upwind and downwind leg, drift should not be encountered. However, it may be difficult to find a situation where the wind is blowing exactly parallel to the field boundaries. This makes it necessary to use a slight wind correction angle on all the legs. It is important to anticipate the turns to compensate for groundspeed, drift, and turning radius. When the wind is behind the gyroplane, the turn must be faster and steeper; when it is ahead of the gyroplane, the turn must be slower and shallower. These same techniques apply while flying in an airport traffic pattern.

S-TURNS

Another training maneuver you might use is the S-turn, which helps you correct for wind drift in turns. This maneuver requires turns to the left and right. The reference line used, whether a road, railroad, or fence, should be straight for a considerable distance and should extend as nearly perpendicular to the wind as possible.

The object of S-turns is to fly a pattern of two half circles of equal size on opposite sides of the reference line. [Figure 20-11] The maneuver should be performed at a constant altitude of 600 to 1,000 feet above the terrain. S-turns may be started at any point; however, during early training it may be beneficial to start on a downwind heading. Entering downwind permits the immediate selection of the steepest bank

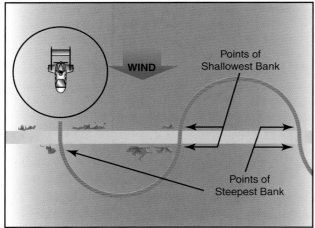

Figure 20-11. S-turns across a road.

that is desired throughout the maneuver. The discussion that follows is based on choosing a reference line that is perpendicular to the wind and starting the maneuver on a downwind heading.

As the gyroplane crosses the reference line, immediately establish a bank. This initial bank is the steepest

used throughout the maneuver since the gyroplane is headed directly downwind and the groundspeed is at its highest. Gradually reduce the bank, as necessary, to describe a ground track of a half circle. Time the turn so that as the rollout is completed, the gyroplane is crossing the reference line perpendicular to it and heading directly upwind. Immediately enter a bank in the opposite direction to begin the second half of the "S." Since the gyroplane is now on an upwind heading, this bank (and the one just completed before crossing the reference line) is the shallowest in the maneuver. Gradually increase the bank, as necessary, to describe a ground track that is a half circle identical in size to the one previously completed on the other side of the reference line. The steepest bank in this turn should be attained just prior to rollout when the gyroplane is approaching the reference line nearest the downwind heading. Time the turn so that as the rollout is complete, the gyroplane is perpendicular to the reference line and is again heading directly downwind.

In summary, the angle of bank required at any given point in the maneuver is dependent on the groundspeed. The faster the groundspeed, the steeper the bank; the slower the groundspeed, the shallower the bank. To express it another way, the more nearly the gyroplane is to a downwind heading, the steeper the bank; the more nearly it is to an upwind heading, the shallower the bank. In addition to varying the angle of bank to correct for drift in order to maintain the proper radius of turn, the gyroplane must also be flown with a drift correction angle (crab) in relation to its ground track; except of course, when it is on direct upwind or downwind headings or there is no wind. One would normally think of the fore and aft axis of the gyroplane as being tangent to the ground track pattern at each point. However, this is not the case. During the turn on the upwind side of the reference line (side from which the wind is blowing), crab the nose of the gyroplane toward the outside of the circle. During the turn on the downwind side of the reference line (side of the reference line opposite to the direction from which the wind is blowing), crab the nose of the gyroplane toward the inside of the circle. In either case, it is obvious that the gyroplane is being crabbed into the wind just as it is when trying to maintain a straight ground track. The amount of crab depends upon the wind velocity and how nearly the gyroplane is to a crosswind position. The stronger the wind, the greater the crab angle at any given position for a turn of a given radius. The more nearly the gyroplane is to a crosswind position, the greater the crab angle. The maximum crab angle should be at the point of each half circle farthest from the reference line.

A standard radius for S-turns cannot be specified, since the radius depends on the airspeed of the gyroplane, the

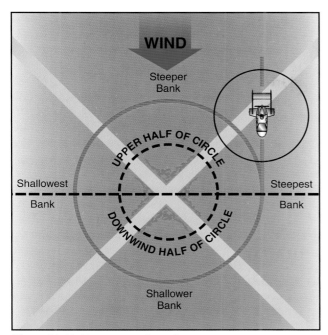

Figure 20-12. Turns around a point.

velocity of the wind, and the initial bank chosen for entry.

TURNS AROUND A POINT

This training maneuver requires you to fly constant radius turns around a preselected point on the ground using a maximum bank of approximately 40°, while maintaining a constant altitude. [Figure 20-12] Your objective, as in other ground reference maneuvers, is to develop the ability to subconsciously control the gyroplane while dividing attention between the flight path and ground references, while still watching for other air traffic in the vicinity.

The factors and principles of drift correction that are involved in S-turns are also applicable in this maneuver. As in other ground track maneuvers, a constant radius around a point will, if any wind exists, require a constantly changing angle of bank and angles of wind correction. The closer the gyroplane is to a direct downwind heading where the groundspeed is greatest, the steeper the bank, and the faster the rate of turn required to establish the proper wind correction angle. The more nearly it is to a direct upwind heading where the groundspeed is least, the shallower the bank, and the slower the rate of turn required to establish the proper wind correction angle. It follows then, that throughout the maneuver, the bank and rate of turn must be gradually varied in proportion to the groundspeed.

The point selected for turns around a point should be prominent and easily distinguishable, yet small enough to present a precise reference. Isolated trees, crossroads, or other similar small landmarks are usually suitable. The point should be in an area away from communities, livestock, or groups of people on the ground to prevent possible annoyance or hazard to others. Since the maneuver is performed between 600 and 1,000 feet AGL, the area selected should also afford an opportunity for a safe emergency landing in the event it becomes necessary.

To enter turns around a point, fly the gyroplane on a downwind heading to one side of the selected point at a distance equal to the desired radius of turn. When any significant wind exists, it is necessary to roll into the initial bank at a rapid rate so that the steepest bank is attained abeam the point when the gyroplane is headed directly downwind. By entering the maneuver while heading directly downwind, the steepest bank can be attained immediately. Thus, if a bank of 40° is desired, the initial bank is 40° if the gyroplane is at the correct distance from the point. Thereafter, the bank is gradually shallowed until the point is reached where the gyroplane is headed directly upwind. At this point, the bank is gradually steepened until the steepest bank is again attained when heading downwind at the initial point of entry.

Just as S-turns require that the gyroplane be turned into the wind, in addition to varying the bank, so do turns around a point. During the downwind half of the circle, the gyroplane's nose must be progressively turned toward the inside of the circle; during the upwind half, the nose must be progressively turned toward the outside. The downwind half of the turn around the point may be compared to the downwind side of the S-turn, while the upwind half of the turn around a point may be compared to the upwind side of the S-turn.

As you become experienced in performing turns around a point and have a good understanding of the effects of wind drift and varying of the bank angle and wind correction angle, as required, entry into the maneuver may be from any point. When entering this maneuver at any point, the radius of the turn must be carefully selected, taking into account the wind velocity and groundspeed, so that an excessive bank is not required later on to maintain the proper ground track.

COMMON ERRORS DURING GROUND REFERENCE MANEUVERS

1. Faulty entry technique.

2. Poor planning, orientation, or division of attention.

3. Uncoordinated flight control application.

4. Improper correction for wind drift.

5. An unsymmetrical ground track during S-turns across a road.

6. Failure to maintain selected altitude or airspeed.

7. Selection of a ground reference where there is no suitable emergency landing site.

FLIGHT AT SLOW AIRSPEEDS

The purpose of maneuvering during slow flight is to help you develop a feel for controlling the gyroplane at slow airspeeds, as well as gain an understanding of how load factor, pitch attitude, airspeed, and altitude control relate to each other.

Like airplanes, gyroplanes have a specific amount of power that is required for flight at various airspeeds, and a fixed amount of power available from the engine. This data can be charted in a graph format. [Figure 20-13] The lowest point of the power required curve represents the speed at which the gyroplane will fly in level flight while using the least amount of power. To fly faster than this speed, or slower, requires more power. While practicing slow flight in a gyroplane, you will likely be operating in the performance realm on the chart that is left of the minimum power required speed. This is often referred to as the "backside of the power curve," or flying "behind the power curve." At these speeds, as pitch is increased to slow the gyroplane, more and more power is required to maintain level flight. At the point where maximum power available is being used, no further reduction in airspeed is possible without initiating a descent. This speed is referred to as the minimum level flight speed. Because there is no excess power available for acceleration, recovery from minimum level flight speed requires lowering the nose of the gyroplane and using altitude to regain airspeed. For this reason, it is essential to practice slow flight at altitudes that allow sufficient height for a safe recovery. Unintentionally flying a gyroplane on the backside of the power curve during approach and landing can be extremely hazardous. Should a go-around become necessary, sufficient altitude to regain airspeed and initiate a climb may not be available, and ground contact may be unavoidable.

Flight at slow airspeeds is usually conducted at airspeeds 5 to 10 m.p.h. above the minimum level flight airspeed. When flying at slow airspeeds, it is important that your control inputs be smooth and slow to prevent a rapid loss of airspeed due to the high drag increases with small changes in pitch attitude. In addition, turns should be limited to shallow bank angles. In order to prevent losing altitude during turns, power must be added. Directional control remains very good while flying at slow airspeeds, because of the high velocity slipstream produced by the increased engine power.

Recovery to cruise flight speed is made by lowering the nose and increasing power. When the desired speed is reached, reduce power to the normal cruise power setting.

COMMON ERRORS

1. Improper entry technique.

2. Failure to establish and maintain an appropriate airspeed.

3. Excessive variations of altitude and heading when a constant altitude and heading are specified.

4. Use of too steep a bank angle.

5. Rough or uncoordinated control technique.

HIGH RATE OF DESCENT

A gyroplane will descend at a high rate when flown at very low forward airspeeds. This maneuver may be entered intentionally when a steep descent is desired, and can be performed with or without power. An unintentional high rate of descent can also occur as a result

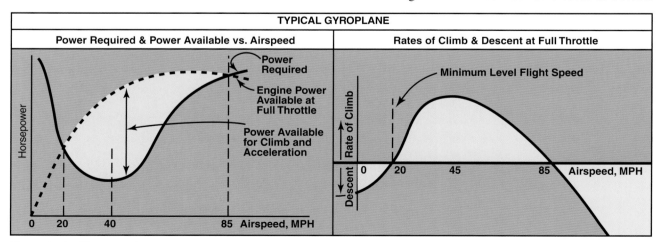

Figure 20-13. The low point on the power required curve is the speed that the gyroplane can fly while using the least amount of power, and is also the speed that will result in a minimum sink rate in a power-off glide.

of failing to monitor and maintain proper airspeed. In powered flight, if the gyroplane is flown below minimum level flight speed, a descent results even though full engine power is applied. Further reducing the airspeed with aft cyclic increases the rate of descent. For gyroplanes with a high thrust-to-weight ratio, this maneuver creates a very high pitch attitude. To recover, the nose of the gyroplane must lowered slightly to exchange altitude for an increase in airspeed.

When operating a gyroplane in an unpowered glide, slowing to below the best glide speed can also result in a high rate of descent. As airspeed decreases, the rate of descent increases, reaching the highest rate as forward speed approaches zero. At slow airspeeds without the engine running, there is very little airflow over the tail surfaces and rudder effectiveness is greatly reduced. Rudder pedal inputs must be exaggerated to maintain effective yaw control. To recover, add power, if available, or lower the nose and allow the gyroplane to accelerate to the proper airspeed. This maneuver demonstrates the importance of maintaining the proper glide speed during an engine-out emergency landing. Attempting to stretch the glide by raising the nose results in a higher rate of descent at a lower forward speed, leaving less distance available for the selection of a landing site.

COMMON ERRORS

1. Improper entry technique.

2. Failure to recognize a high rate of descent.

3. Improper use of controls during recovery.

4. Initiation of recovery below minimum recovery altitude.

LANDINGS

Landings may be classified according to the landing surface, obstructions, and atmospheric conditions. Each type of landing assumes that certain conditions exist. To meet the actual conditions, a combination of techniques may be necessary.

NORMAL LANDING

The procedure for a normal landing in a gyroplane is predicated on having a prepared landing surface and no significant obstructions in the immediate area. After entering a traffic pattern that conforms to established standards for the airport and avoids the flow of fixed wing traffic, a before landing checklist should be reviewed. The extent of the items on the checklist is dependent on the complexity of the gyroplane, and can include fuel, mixture, carburetor heat, propeller, engine instruments, and a check for traffic.

Gyroplanes experience a slight lag between control input and aircraft response. This lag becomes more apparent during the sensitive maneuvering required for landing, and care must be taken to avoid overcorrecting for deviations from the desired approach path. After the turn to final, the approach airspeed appropriate for the gyroplane should be established. This speed is normally just below the minimum power required speed for the gyroplane in level flight. During the approach, maintain this airspeed by making adjustments to the gyroplane's pitch attitude, as necessary. Power is used to control the descent rate.

Approximately 10 to 20 feet above the runway, begin the flare by gradually increasing back pressure on the cyclic to reduce speed and decrease the rate of descent. The gyroplane should reach a near-zero rate of descent approximately 1 foot above the runway with the power at idle. Low airspeed combined with a minimum of propwash over the tail surfaces reduces rudder effectiveness during the flare. If a yaw moment is encountered, use whatever rudder control is required to maintain the desired heading. The gyroplane should be kept laterally level and with the longitudinal axis in the direction of ground track. Landing with sideward motion can damage the landing gear and must be avoided. In a full-flare landing, attempt to hold the gyroplane just off the runway by steadily increasing back pressure on the cyclic. This causes the gyroplane to settle slowly to the runway in a slightly nose-high attitude as forward momentum dissipates.

Ground roll for a full-flare landing is typically under 50 feet, and touchdown speed under 20 m.p.h. If a 20 m.p.h. or greater headwind exists, it may be necessary to decrease the length of the flare and allow the gyroplane to touch down at a slightly higher airspeed to prevent it from rolling backward on landing. After touchdown, rotor r.p.m. decays rather rapidly. On landings where brakes are required immediately after touchdown, apply them lightly, as the rotor is still carrying much of the weight of the aircraft and too much braking causes the tires to skid.

SHORT-FIELD LANDING

A short-field landing is necessary when you have a relatively short landing area or when an approach must be made over obstacles that limit the available landing area. When practicing short-field landings, assume you are making the approach and landing over a 50-foot obstruction in the approach area.

To conduct a short-field approach and landing, follow normal procedures until you are established on the final approach segment. At this point, use aft cyclic to reduce airspeed below the speed for minimum sink. By decreasing speed, sink rate increases and a steeper approach path is achieved, minimizing the distance between clearing the obstacle and making contact with the surface. [Figure 20-14] The approach speed must remain fast enough, however, to allow the flare to arrest the forward and vertical speed of the gyroplane. If the approach speed is too low, the remaining vertical momentum will result in a hard landing. On a short-field landing with a slight headwind, a touchdown with no ground roll is possible. Without wind, the ground roll is normally less than 50 feet.

SOFT-FIELD LANDING
Use the soft-field landing technique when the landing surface presents high wheel drag, such as mud, snow, sand, tall grass or standing water. The objective is to transfer the weight of the gyroplane from the rotor to the landing gear as gently and slowly as possible. With a headwind close to the touchdown speed of the gyroplane, a power approach can be made close to the minimum level flight speed. As you increase the nose pitch attitude just prior to touchdown, add additional power to cushion the landing. However, power should be removed, just as the wheels are ready to touch. This results is a very slow, gentle touchdown. In a strong headwind, avoid allowing the gyroplane to roll rearward at touchdown. After touchdown, smoothly and gently lower the nosewheel to the ground. Minimize the use of brakes, and remain aware that the nosewheel could dig in the soft surface.

When no wind exists, use a steep approach similar to a short-field landing so that the forward speed can be dissipated during the flare. Use the throttle to cushion the touchdown.

CROSSWIND LANDING
Crosswind landing technique is normally used in gyroplanes when a crosswind of approximately 15 m.p.h. or less exists. In conditions with higher crosswinds, it becomes very difficult, if not impossible, to maintain adequate compensation for the crosswind. In these conditions, the slow touchdown speed of a gyroplane allows a much safer option of turning directly into the wind and landing with little or no ground roll. Deciding when to use this technique, however, may be complicated by gusting winds or the characteristics of the particular landing area.

On final approach, establish a crab angle into the wind to maintain a ground track that is aligned with the extended centerline of the runway. Just before touchdown, remove the crab angle and bank the gyroplane slightly into the wind to prevent drift. Maintain longitudinal alignment with the runway using the rudder. In higher crosswinds, if full rudder deflection is not sufficient to maintain alignment with the runway, applying a slight amount of power can increase rudder effectiveness. The length of the flare should be reduced to allow a slightly higher touchdown speed than that used in a no-wind landing. Touchdown is made on the upwind main wheel first, with the other main wheel settling to the runway as forward momentum is lost. After landing, continue to keep the rotor tilted into the wind to maintain positive control during the rollout.

HIGH-ALTITUDE LANDING
A high-altitude landing assumes a density altitude near the limit of what is considered good climb performance

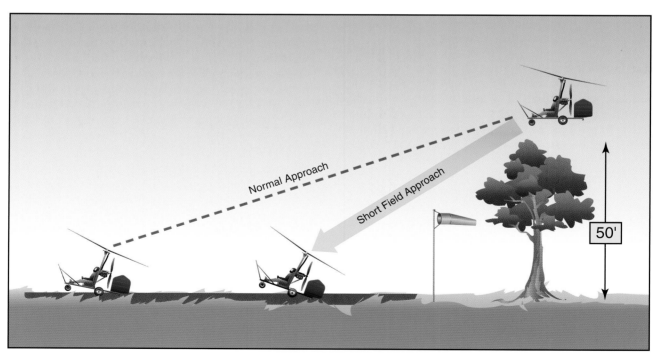

Figure 20-14. The airspeed used on a short-field approach is slower than that for a normal approach, allowing a steeper approach path and requiring less runway.

for the gyroplane. When using the same indicated airspeed as that used for a normal approach at lower altitude, a high density altitude results in higher rotor r.p.m. and a slightly higher rate of descent. The greater vertical velocity is a result of higher true airspeed as compared with that at low altitudes. When practicing high-altitude landings, it is prudent to first learn normal landings with a flare and roll out. Full flare, no roll landings should not be attempted until a good feel for aircraft response at higher altitudes has been acquired. As with high-altitude takeoffs, it is also important to consider the effects of higher altitude on engine performance.

COMMON ERRORS DURING LANDING

1. Failure to establish and maintain a stabilized approach.

2. Improper technique in the use of power.

3. Improper technique during flare or touchdown.

4. Touchdown at too low an airspeed with strong headwinds, causing a rearward roll.

5. Poor directional control after touchdown.

6. Improper use of brakes.

GO-AROUND

The go-around is used to abort a landing approach when unsafe factors for landing are recognized. If the decision is made early in the approach to go around, normal climb procedures utilizing V_X and V_Y should be used. A late decision to go around, such as after the full flare has been initiated, may result in an airspeed where power required is greater than power available. When this occurs, a touchdown becomes unavoidable and it may be safer to proceed with the landing than to sustain an extended ground roll that would be required

to go around. Also, the pitch attitude of the gyroplane in the flare is high enough that the tail would be considerably lower than the main gear, and a touch down with power on would result in a sudden pitch down and acceleration of the aircraft. Control of the gyroplane under these circumstances may be difficult. Consequently, the decision to go around should be made as early as possible, before the speed is reduced below the point that power required exceeds power available.

COMMON ERRORS

1. Failure to recognize a situation where a go-around is necessary.

2. Improper application of power.

3. Failure to control pitch attitude.

4. Failure to maintain recommended airspeeds.

5. Failure to maintain proper track during climb out.

AFTER LANDING AND SECURING

The after-landing checklist should include such items as the transponder, cowl flaps, fuel pumps, lights, and magneto checks, when so equipped. The rotor blades demand special consideration after landing, as turning rotor blades can be hazardous to others. Never enter an area where people or obstructions are present with the rotor turning. To assist the rotor in slowing, tilt the cyclic control into the prevailing wind or face the gyroplane downwind. When slowed to under approximately 75 r.p.m., the rotor brake may be applied, if available. Use caution as the rotor slows, as excess taxi speed or high winds could cause blade flap to occur. The blades should be depitched when taxiing if a collective control is available. When leaving the gyroplane, always secure the blades with a tiedown or rotor brake.

Gyroplanes are quite reliable, however emergencies do occur, whether a result of mechanical failure or pilot error. By having a thorough knowledge of the gyroplane and its systems, you will be able to more readily handle the situation. In addition, by knowing the conditions which can lead to an emergency, many potential accidents can be avoided.

ABORTED TAKEOFF

Prior to every takeoff, consideration must be given to a course of action should the takeoff become undesirable or unsafe. Mechanical failures, obstructions on the takeoff surface, and changing weather conditions are all factors that could compromise the safety of a takeoff and constitute a reason to abort. The decision to abort a takeoff should be definitive and made as soon as an unsafe condition is recognized. By initiating the abort procedures early, more time and distance will be available to bring the gyroplane to a stop. A late decision to abort, or waiting to see if it will be necessary to abort, can result in a dangerous situation with little time to respond and very few options available.

When initiating the abort sequence prior to the gyroplane leaving the surface, the procedure is quite simple. Reduce the throttle to idle and allow the gyroplane to decelerate, while slowly applying aft cyclic for aerodynamic braking. This technique provides the most effective braking and slows the aircraft very quickly. If the gyroplane has left the surface when the decision to abort is made, reduce the throttle until an appropriate descent rate is achieved. Once contact with the surface is made, reduce the throttle to idle and apply aerodynamic braking as before. The wheel brakes, if the gyroplane is so equipped, may be applied, as necessary, to assist in slowing the aircraft.

ACCELERATE/STOP DISTANCE

An accelerate/stop distance is the length of ground roll an aircraft would require to accelerate to takeoff speed and, assuming a decision to abort the takeoff is made, bring the aircraft safely to a stop. This value changes for a given aircraft based on atmospheric conditions, the takeoff surface, aircraft weight, and other factors affecting performance. Knowing the accelerate/stop value for your gyroplane can be helpful in planning a

safe takeoff, but having this distance available does not necessarily guarantee a safe aborted takeoff is possible for every situation. If the decision to abort is made after liftoff, for example, the gyroplane will require considerably more distance to stop than the accelerate/stop figure, which only considers the ground roll requirement. Planning a course of action for an abort decision at various stages of the takeoff is the best way to ensure the gyroplane can be brought safely to a stop should the need arise.

For a gyroplane without a flight manual or other published performance data, the accelerate/stop distance can be reasonably estimated once you are familiar with the performance and takeoff characteristics of the aircraft. For a more accurate figure, you can accelerate the gyroplane to takeoff speed, then slow to a stop, and note the distance used. Doing this several times gives you an average accelerate/stop distance. When performance charts for the aircraft are available, as in the flight manual of a certificated gyroplane, accurate accelerate/stop distances under various conditions can be determined by referring to the ground roll information contained in the charts.

LIFT-OFF AT LOW AIRSPEED AND HIGH ANGLE OF ATTACK

Because of ground effect, your gyroplane might be able to become airborne at an airspeed less than minimum level flight speed. In this situation, the gyroplane is flying well behind the power curve and at such a high angle of attack that unless a correction is made, there will be little or no acceleration toward best climb speed. This condition is often encountered in gyroplanes capable of jump takeoffs. Jumping without sufficient rotor inertia to allow enough time to accelerate through minimum level flight speed, usually results in your gyroplane touching down after liftoff. If you do touch down after performing a jump takeoff, you should abort the takeoff.

During a rolling takeoff, if the gyroplane is forced into the air too early, you could get into the same situation. It is important to recognize this situation and take immediate corrective action. You can either abort the takeoff, if enough runway exists, or lower the nose and accelerate to the best climb speed. If you choose to continue the takeoff, verify that full power is applied, then, slowly lower the nose, making sure the gyroplane does not contact the surface. While in ground effect, accelerate to the best climb speed. Then, adjust the nose pitch attitude to maintain that airspeed.

COMMON ERRORS
The following errors might occur when practicing a lift-off at a low airspeed.

1. Failure to check rotor for proper operation, track, and r.p.m. prior to initiating takeoff.

2. Use of a power setting that does not simulate a "behind the power curve" situation.

3. Poor directional control.

4. Rotation at a speed that is inappropriate for the maneuver.

5. Poor judgement in determining whether to abort or continue takeoff.

6. Failure to establish and maintain proper climb attitude and airspeed, if takeoff is continued.

7. Not maintaining the desired ground track during the climb.

PILOT-INDUCED OSCILLATION (PIO)
Pilot-induced oscillation, sometimes referred to as porpoising, is an unintentional up-and-down oscillation of the gyroplane accompanied with alternating climbs and descents of the aircraft. PIO is often the result of an inexperienced pilot overcontrolling the gyroplane, but this condition can also be induced by gusty wind conditions. While this condition is usually thought of as a longitudinal problem, it can also happen laterally.

As with most other rotor-wing aircraft, gyroplanes experience a slight delay between control input and the reaction of the aircraft. This delay may cause an inexperienced pilot to apply more control input than required, causing a greater aircraft response than was desired. Once the error has been recognized, opposite control input is applied to correct the flight attitude. Because of the nature of the delay in aircraft response, it is possible for the corrections to be out of synchronization with the movements of the aircraft and aggravate the undesired changes in attitude. The result is PIO, or unintentional oscillations that can grow rapidly in magnitude. [Figure 21-1]

In gyroplanes with an open cockpit and limited flight instruments, it can be difficult for an inexperienced pilot to recognize a level flight attitude due to the lack of visual references. As a result, PIO can develop as the pilot chases a level flight attitude and introduces climbing and descending oscillations. PIO can also develop if a wind gust displaces the aircraft, and the control inputs made to correct the attitude are out of phase with the aircraft movements. Because the rotor disc angle decreases at higher speeds and cyclic control becomes more sensitive, PIO is more likely to occur and can be more pronounced at high airspeeds. To minimize the possibility of PIO, avoid high-speed flight in gusty conditions, and make only small control inputs. After making a control input, wait briefly and observe the reaction of the aircraft before making another input. If PIO is encountered, reduce power and place the cyclic in the position for a normal climb. Once the oscillations have stopped, slowly return the throttle and cyclic to their normal positions. The likelihood of encountering PIO decreases greatly as experience is gained, and the ability to subconsciously anticipate the reactions of the gyroplane to control inputs is developed.

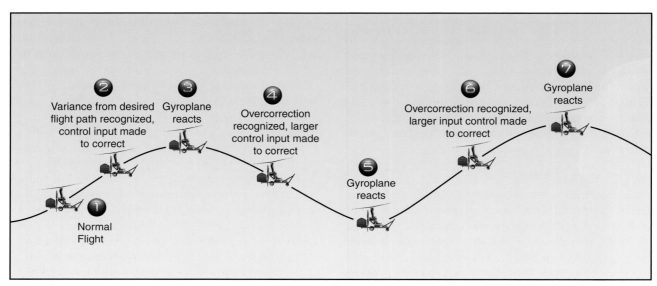

Figure 21-1. Pilot-induced oscillation can result if the gyroplane's reactions to control inputs are not anticipated and become out of phase.

BUNTOVER (POWER PUSHOVER)

As you learned in Chapter 16—Gyroplane Aerodynamics, the stability of a gyroplane is greatly influenced by rotor force. If rotor force is rapidly removed, some gyroplanes have a tendency to pitch forward abruptly. This is often referred to as a forward tumble, buntover, or power pushover. Removing the rotor force is often referred to as unloading the rotor, and can occur if pilot-induced oscillations become excessive, if extremely turbulent conditions are encountered, or the nose of the gyroplane is pushed forward rapidly after a steep climb.

A power pushover can occur on some gyroplanes that have the propeller thrust line above the center of gravity and do not have an adequate horizontal stabilizer. In this case, when the rotor is unloaded, the propeller thrust magnifies the pitching moment around the center of gravity. Unless a correction is made, this nose pitching action could become self-sustaining and irreversible. An adequate horizontal stabilizer slows the pitching rate and allows time for recovery.

Since there is some disagreement between manufacturers as to the proper recovery procedure for this situation, you must check with the manufacturer of your gyroplane. In most cases, you need to remove power and load the rotor blades. Some manufacturers, especially those with gyroplanes where the propeller thrust line is above the center of gravity, recommend that you need to immediately remove power in order to prevent a power pushover situation. Other manufacturers recommend that you first try to load the rotor blades. For the proper positioning of the cyclic when loading up the rotor blades, check with the manufacturer.

When compared to other aircraft, the gyroplane is just as safe and very reliable. The most important factor, as in all aircraft, is pilot proficiency. Proper training and flight experience helps prevent the risks associated with pilot-induced oscillation or buntover.

GROUND RESONANCE

Ground resonance is a potentially damaging aerodynamic phenomenon associated with articulated rotor systems. It develops when the rotor blades move out of phase with each other and cause the rotor disc to become unbalanced. If not corrected, ground resonance can cause serious damage in a matter of seconds.

Ground resonance can only occur while the gyroplane is on the ground. If a shock is transmitted to the rotor system, such as with a hard landing on one gear or when operating on rough terrain, one or more of the blades could lag or lead and allow the rotor system's center of gravity to be displaced from the center of rotation. Subsequent shocks to the other gear aggravate the imbalance causing the rotor center of gravity to rotate around the hub. This phenomenon is not unlike an out-of-balance washing machine. [Figure 21-2]

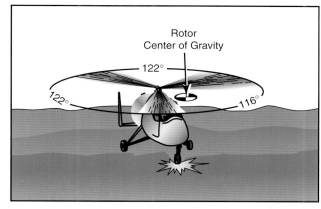

Figure 21-2. Taxiing on rough terrain can send a shock wave to the rotor system, resulting in the blades of a three-bladed rotor system moving from their normal 120° relationship to each other.

To reduce the chance of experiencing ground resonance, every preflight should include a check for proper strut inflation, tire pressure, and lag-lead damper operation. Improper strut or tire inflation can change the vibration frequency of the airframe, while improper damper settings change the vibration frequency of the rotor.

If you experience ground resonance, and the rotor r.p.m. is not yet sufficient for flight, apply the rotor brake to maximum and stop the rotor as soon as possible. If ground resonance occurs during takeoff, when rotor r.p.m. is sufficient for flight, lift off immediately. Ground resonance cannot occur in flight, and the rotor blades will automatically realign themselves once the gyroplane is airborne. When prerotating the rotor system prior to takeoff, a slight vibration may be felt that is a very mild form of ground resonance. Should this oscillation amplify, discontinue the prerotation and apply maximum rotor brake.

EMERGENCY APPROACH AND LANDING

The modern engines used for powering gyroplanes are generally very reliable, and an actual mechanical malfunction forcing a landing is not a common occurrence. Failures are possible, which necessitates planning for and practicing emergency approaches and landings. The best way to ensure that important items are not overlooked during an emergency procedure is to use a checklist, if one is available and time permits. Most gyroplanes do not have complex electrical, hydraulic, or pneumatic systems that require lengthy checklists. In these aircraft, the checklist can be easily committed to memory so that immediate action can be taken if needed. In addition, you should always maintain an awareness of your surroundings and be constantly on the alert for suitable emergency landing sites.

When an engine failure occurs at altitude, the first course of action is to adjust the gyroplane's pitch attitude to achieve the best glide speed. This yields the most distance available for a given altitude, which in turn, allows for more possible landing sites. A common mistake when learning emergency procedures is attempting to stretch the glide by raising the nose, which instead results in a steep approach path at a slow airspeed and a high rate of descent. [Figure 21-3] Once you have attained best glide speed, scan the area within gliding distance for a suitable landing site. Remember to look behind the aircraft, as well as in front, making gentle turns, if necessary, to see around the airframe. When selecting a landing site, you must consider the wind direction and speed, the size of the landing site, obstructions to the approach, and the condition of the surface. A site that allows a landing into the wind and has a firm, smooth surface with no obstructions is the most desirable. When considering landing on a road, be alert for powerlines, signs, and automobile traffic. In many cases, an ideal site will not be available, and it will be necessary for you to evaluate your options and choose the best alternative. For example, if a steady wind will allow a touchdown with no ground roll, it may be acceptable to land in a softer field or in a

smaller area than would normally be considered. On landing, use short or soft field technique, as appropriate, for the site selected. A slightly higher-than-normal approach airspeed may be required to maintain adequate airflow over the rudder for proper yaw control.

EMERGENCY EQUIPMENT AND SURVIVAL GEAR

On any flight not in the vicinity of an airport, it is highly advisable to prepare a survival kit with items that would be necessary in the event of an emergency. A properly equipped survival kit should be able to provide you with sustenance, shelter, medical care, and a means to summon help without a great deal of effort on your part. An efficient way to organize your survival kit is to prepare a basic core of supplies that would be necessary for any emergency, and allow additional space for supplementary items appropriate for the terrain and weather you expect for a particular flight. The basic items to form the basis of your survival kit would typically include: a first-aid kit and field medical guide, a flashlight, water, a knife, matches, some type of shelter, and a signaling device. Additional items that may be added to meet the conditions, for example, would be a lifevest for a flight over water, or heavy clothing for a flight into cold weather. Another consideration is carrying a cellular phone. Several pilots have been rescued after calling someone to indicate there had been an accident.

Figure 21-3. Any deviation from best glide speed will reduce the distance you can glide and may cause you to land short of a safe touchdown point.

Aeronautical Decision Making

As with any aircraft, the ability to pilot a gyroplane safely is largely dependent on the capacity of the pilot to make sound and informed decisions. To this end, techniques have been developed to ensure that a pilot uses a systematic approach to making decisions, and that the course of action selected is the most appropriate for the situation. In addition, it is essential that you learn to evaluate your own fitness, just as you evaluate the airworthiness of your aircraft, to ensure that your physical and mental condition is compatible with a safe flight. The techniques for acquiring these essential skills are explained in depth in Chapter 14—Aeronautical Decision Making (Helicopter).

As explained in Chapter 14, one of the best methods to develop your aeronautical decision making is learning to recognize the five hazardous attitudes, and how to counteract these attitudes. [Figure 22-1] This chapter focuses on some examples of how these hazardous attitudes can apply to gyroplane operations.

HAZARDOUS ATTITUDE	ANTIDOTE
Impulsivity: "Do something—quickly!"	"Not so fast. Think first."
Invulnerability: "It won't happen to me!"	"It could happen to me."
Macho: "I can do it."	"Taking chances is foolish."
Resignation: "What's the use?"	"I'm not helpless. I can make the difference."
Anti-authority: "Don't tell me!"	"Follow the rules. They are usually right."

Figure 22-1. To overcome hazardous attitudes, you must memorize the antidotes for each of them. You should know them so well that they will automatically come to mind when you need them.

IMPULSIVITY

Gyroplanes are a class of aircraft which can be acquired, constructed, and operated in ways unlike most other aircraft. This inspires some of the most exciting and rewarding aspects of flying, but it also creates a unique set of dangers to which a gyroplane pilot must be alert. For example, a wide variety of amateur-built gyroplanes are available, which can be purchased in kit form and assembled at home. This makes the airworthiness of these gyroplanes ultimately dependent on the vigilance of the one assembling and maintaining the aircraft. Consider the following scenario.

Jerry recently attended an airshow that had a gyroplane flight demonstration and a number of gyroplanes on display. Being somewhat mechanically inclined and retired with available spare time, Jerry decided that building a gyroplane would be an excellent project for him and ordered a kit that day. When the kit arrived, Jerry unpacked it in his garage and immediately began the assembly. As the gyroplane neared completion, Jerry grew more excited at the prospect of flying an aircraft that he had built with his own hands. When the gyroplane was nearly complete, Jerry noticed that a rudder cable was missing from the kit, or perhaps lost during the assembly. Rather than contacting the manufacturer and ordering a replacement, which Jerry thought would be a hassle and too time consuming, he went to his local hardware store and purchased some cable he thought would work. Upon returning home, he was able to fashion a rudder cable that seemed functional and continued with the assembly.

Jerry is exhibiting "impulsivity." Rather than taking the time to properly build his gyroplane to the specifications set forth by the manufacturer, Jerry let his excitement allow him to cut corners by acting on impulse, rather than taking the time to think the matter through. Although some enthusiasm is normal during assembly, it should not be permitted to compromise the airworthiness of the aircraft. Manufacturers often use high quality components, which are constructed and tested to standards much higher than those found in hardware stores. This is particularly true in the area of cables, bolts, nuts, and other types of fasteners where strength is essential. The proper course of action Jerry should have taken would be to stop, think, and consider the possible consequences of making an impulsive decision. Had he realized that a broken rudder cable in flight could cause a loss of control of the gyroplane, he likely would have taken the time to contact the manufacturer and order a cable that met the design specifications.

INVULNERABILITY

Another area that can often lead to trouble for a gyroplane pilots is the failure to obtain adequate flight instruction to operate their gyroplane safely. This can be the result of people thinking that because they can build the machine themselves, it must be simple enough to learn how to fly by themselves. Other reasons that can lead to this problem can be simply monetary, in not wanting to pay the money for adequate instruction, or feeling that because they are qualified in another type of aircraft, flight instruction is not necessary. In reality, gyroplane operations are quite unique, and there is no substitute for adequate training by a competent and authorized instructor. Consider the following scenario.

Jim recently met a coworker who is a certified pilot and owner of a two-seat gyroplane. In discussing the gyroplane with his coworker, Jim was fascinated and reminded of his days in the military as a helicopter pilot many years earlier. When offered a ride, Jim readily accepted. He met his coworker at the airport the following weekend for a short flight and was immediately hooked. After spending several weeks researching available designs, Jim decided on a particular gyroplane and purchased a kit. He had it assembled in a few months, with the help and advice of his new friend and fellow gyroplane enthusiast. When the gyroplane was finally finished, Jim asked his friend to take him for a ride in his two-seater to teach him the basics of flying. The rest, he said, he would figure out while flying his own machine from a landing strip that he had fashioned in a field behind his house.

Jim is unknowingly inviting disaster by allowing himself to be influenced by the hazardous attitude of "invulnerability." Jim does not feel that it is possible to have an accident, probably because of his past experience in helicopters and from witnessing the ease with which his coworker controlled the gyroplane on their flight together. What Jim is failing to consider, however, is the amount of time that has passed since he was proficient in helicopters, and the significant differences between helicopter and gyroplane operations. He is also overlooking the fact that his friend is a certificated pilot, who has taken a considerable amount of instruction to reach his level of competence. Without adequate instruction and experience, Jim could, for example, find himself in a pilot-induced oscillation without knowing the proper technique for recovery, which could ultimately be disastrous. The antidote for an attitude of invulnerability is to realize that accidents can happen to anyone.

MACHO

Due to their unique design, gyroplanes are quite responsive and have distinct capabilities. Although gyroplanes are capable of incredible maneuvers, they do have limitations. As gyroplane pilots grow more comfortable with their machines, they might be tempted to operate progressively closer to the edge of the safe operating envelope. Consider the following scenario.

Pat has been flying gyroplanes for years and has an excellent reputation as a skilled pilot. He has recently built a high performance gyroplane with an advanced rotor system. Pat was excited to move into a more advanced aircraft because he had seen the same design performing aerobatics in an airshow earlier that year. He was amazed by the capability of the machine. He had always felt that his ability surpassed the capability of the aircraft he was flying. He had invested a large amount of time and resources into the construction of the aircraft, and, as he neared completion of the assembly, he was excited about the opportunity of showing his friends and family his capabilities.

During the first few flights, Pat was not completely comfortable in the new aircraft, but he felt that he was progressing through the transition at a much faster pace than the average pilot. One morning, when he was with some of his fellow gyroplane enthusiasts, Pat began to brag about the superior handling qualities of the machine he had built. His friends were very excited, and Pat realized that they would be expecting quite a show on his next flight. Not wanting to disappoint them, he decided that although it might be early, he would give the spectators on the ground a real show. On his first pass he came down fairly steep and fast and recovered from the dive with ease. Pat then decided to make another pass only this time he would come in much steeper. As he began to recover, the aircraft did not climb as he expected and almost settled to the ground. Pat narrowly escaped hitting the spectators as he was trying to recover from the dive.

Pat had let the "macho" hazardous attitude influence his decision making. He could have avoided the consequences of this attitude if he had stopped to think that taking chances is foolish.

RESIGNATION

Some of the elements pilots face cannot be controlled. Although we cannot control the weather, we do have some very good tools to help predict what it will do, and how it can affect our ability to fly safely. Good pilots always make decisions that will keep their options open if an unexpected event occurs while flying. One of the greatest resources we have in the cockpit is the ability to improvise and improve the overall situation even when a risk element jeopardizes the probability of a successful flight. Consider the following scenario.

Judi flies her gyroplane out of a small grass strip on her family's ranch. Although the rugged landscape of the ranch lends itself to the remarkable scenery, it leaves few places to safely land in the event of an emergency. The only suitable place to land other than the grass strip is to the west on a smooth section of the road leading to the house. During Judi's training, her traffic patterns were always made with left turns. Figuring this was how she was to make all traffic patterns, she applied this to the grass strip at the ranch. In addition, she was uncomfortable with making turns to the right. Since, the wind at the ranch was predominately from the south, this meant that the traffic pattern was to the east of the strip.

Judi's hazardous attitude is "resignation." She has accepted the fact that her only course of action is to fly east of the strip, and if an emergency happens, there is not much she can do about it. The antidote to this hazardous attitude is "I'm not helpless, I can make a difference." Judi could easily modify her traffic pattern so that she is always within gliding distance of a suitable landing area. In addition, if she was uncomfortable with a maneuver, she could get additional training.

ANTI-AUTHORITY

Regulations are implemented to protect aviation personnel as well as the people who are not involved in aviation. Pilots who choose to operate outside of the regulations, or on the ragged edge, eventually get caught, or even worse, they end up having an accident. Consider the following scenario.

Dick is planning to fly the following morning and realizes that his medical certificate has expired. He knows that he will not have time to take a flight physical before his morning flight. Dick thinks to himself "The rules are too restrictive. Why should I spend the time and money on a physical when I will be the only one at risk if I fly tomorrow?"

Dick decides to fly the next morning thinking that no harm will come as long as no one finds out that he is flying illegally. He pulls his gyroplane out from the hangar, does the preflight inspection, and is getting ready to start the engine when an FAA inspector walks up and greets him. The FAA inspector is conducting a random inspection and asks to see Dick's pilot and medical certificates.

Dick subjected himself to the hazardous attitude of "anti-authority." Now, he will be unable to fly, and has invited an exhaustive review of his operation by the FAA. Dick could have prevented this event if had taken the time to think, "Follow the rules. They are usually right."

22-4

GLOSSARY

ABSOLUTE ALTITUDE—The actual distance an object is above the ground.

ADVANCING BLADE—The blade moving in the same direction as the helicopter or gyroplane. In rotorcraft that have counterclockwise main rotor blade rotation as viewed from above, the advancing blade is in the right half of the rotor disc area during forward movement.

AIRFOIL—Any surface designed to obtain a useful reaction of lift, or negative lift, as it moves through the air.

AGONIC LINE—A line along which there is no magnetic variation.

AIR DENSITY—The density of the air in terms of mass per unit volume. Dense air has more molecules per unit volume than less dense air. The density of air decreases with altitude above the surface of the earth and with increasing temperature.

AIRCRAFT PITCH—When referenced to an aircraft, it is the movement about its lateral, or pitch axis. Movement of the cyclic forward or aft causes the nose of the helicopter or gyroplane to pitch up or down.

AIRCRAFT ROLL—Is the movement of the aircraft about its longitudinal axis. Movement of the cyclic right or left causes the helicopter or gyroplane to tilt in that direction.

AIRWORTHINESS DIRECTIVE —When an unsafe condition exists with an aircraft, the FAA issues an airworthiness directive to notify concerned parties of the condition and to describe the appropriate corrective action.

ALTIMETER—An instrument that indicates flight altitude by sensing pressure changes and displaying altitude in feet or meters.

ANGLE OF ATTACK—The angle between the airfoil's chord line and the relative wind.

ANTITORQUE PEDAL—The pedal used to control the pitch of the tail rotor or air diffuser in a NOTAR® system.

ANTITORQUE ROTOR—See tail rotor.

ARTICULATED ROTOR—A rotor system in which each of the blades is connected to the rotor hub in such a way that it is free to change its pitch angle, and move up and down and fore and aft in its plane of rotation.

AUTOPILOT—Those units and components that furnish a means of automatically controlling the aircraft.

AUTOROTATION—The condition of flight during which the main rotor is driven only by aerodynamic forces with no power from the engine.

AXIS-OF-ROTATION—The imaginary line about which the rotor rotates. It is represented by a line drawn through the center of, and perpendicular to, the tip-path plane.

BASIC EMPTY WEIGHT—The weight of the standard rotorcraft, operational equipment, unusable fuel, and full operating fluids, including full engine oil.

BLADE CONING—An upward sweep of rotor blades as a result of lift and centrifugal force.

BLADE DAMPER—A device attached to the drag hinge to restrain the fore and aft movement of the rotor blade.

BLADE FEATHER OR FEATHERING—The rotation of the blade around the spanwise (pitch change) axis.

BLADE FLAP—The ability of the rotor blade to move in a vertical direction. Blades may flap independently or in unison.

BLADE GRIP—The part of the hub assembly to which the rotor blades are attached, sometimes referred to as blade forks.

BLADE LEAD OR LAG—The fore and aft movement of the blade in the plane of rotation. It is sometimes called hunting or dragging.

BLADE LOADING—The load imposed on rotor blades, determined by dividing the total weight of the helicopter by the combined area of all the rotor blades.

BLADE ROOT—The part of the blade that attaches to the blade grip.

BLADE SPAN—The length of a blade from its tip to its root.

BLADE STALL—The condition of the rotor blade when it is operating at an angle of attack greater than the maximum angle of lift.

BLADE TIP—The further most part of the blade from the hub of the rotor.

BLADE TRACK—The relationship of the blade tips in the plane of rotation. Blades that are in track will move through the same plane of rotation.

BLADE TRACKING—The mechanical procedure used to bring the blades of the rotor into a satisfactory relationship with each other under dynamic conditions so that all blades rotate on a common plane.

BLADE TWIST—The variation in the angle of incidence of a blade between the root and the tip.

BLOWBACK—The tendency of the rotor disc to tilt aft in forward flight as a result of flapping.

BUNTOVER—The tendency of a gyroplane to pitch forward when rotor force is removed.

CALIBRATED AIRSPEED (CAS)—Indicated airspeed of an aircraft, corrected for installation and instrumentation errors.

CENTER OF GRAVITY—The theoretical point where the entire weight of the helicopter is considered to be concentrated.

CENTER OF PRESSURE—The point where the resultant of all the aerodynamic forces acting on an airfoil intersects the chord.

CENTRIFUGAL FORCE—The apparent force that an object moving along a circular path exerts on the body constraining the object and that acts outwardly away from the center of rotation.

CENTRIPETAL FORCE—The force that attracts a body toward its axis of rotation. It is opposite centrifugal force.

CHIP DETECTOR—A warning device that alerts you to any abnormal wear in a transmission or engine. It consists of a magnetic plug located within the transmission. The magnet attracts any metal particles that have come loose from the bearings or other transmission parts. Most chip detectors have warning lights located on the instrument panel that illuminate when metal particles are picked up.

CHORD—An imaginary straight line between the leading and trailing edges of an airfoil section.

CHORDWISE AXIS—A term used in reference to semirigid rotors describing the flapping or teetering axis of the rotor.

COAXIL ROTOR—A rotor system utilizing two rotors turning in opposite directions on the same centerline. This system is used to eliminated the need for a tail rotor.

COLLECTIVE PITCH CONTROL—The control for changing the pitch of all the rotor blades in the main rotor system equally and simultaneously and, consequently, the amount of lift or thrust being generated.

CONING—See blade coning.

CORIOLIS EFFECT—The tendency of a rotor blade to increase or decrease its velocity in its plane of rotation when the center of mass moves closer or further from the axis of rotation.

CYCLIC FEATHERING—The mechanical change of the angle of incidence, or pitch, of individual rotor blades independently of other blades in the system.

CYCLIC PITCH CONTROL—The control for changing the pitch of each rotor blade individually as it rotates through one cycle to govern the tilt of the rotor disc and, consequently, the direction and velocity of horizontal movement.

DELTA HINGE—A flapping hinge with a skewed axis so that the flapping motion introduces a component of feathering that would result in a restoring force in the flap-wise direction.

DENSITY ALTITUDE—Pressure altitude corrected for nonstandard temperature variations.

DEVIATION—A compass error caused by magnetic disturbances from the electrical and metal components in the aircraft. The correction for this error is displayed on a compass correction card place near the magnetic compass of the aircraft.

DIRECT CONTROL—The ability to maneuver a rotorcraft by tilting the rotor disc and changing the pitch of the rotor blades.

DIRECT SHAFT TURBINE—A shaft turbine engine in which the compressor and power section are mounted on a common driveshaft.

DISC AREA—The area swept by the blades of the rotor. It is a circle with its center at the hub and has a radius of one blade length.

DISC LOADING—The total helicopter weight divided by the rotor disc area.

DISSYMMETRY OF LIFT—The unequal lift across the rotor disc resulting from the difference in the velocity of air over the advancing blade half and retreating blade half of the rotor disc area.

DRAG—An aerodynamic force on a body acting parallel and opposite to relative wind.

DUAL ROTOR—A rotor system utilizing two main rotors.

DYNAMIC ROLLOVER—The tendency of a helicopter to continue rolling when the critical angle is exceeded, if one gear is on the ground, and the helicopter is pivoting around that point.

FEATHERING—The action that changes the pitch angle of the rotor blades by rotating them around their feathering (spanwise) axis.

FEATHERING AXIS—The axis about which the pitch angle of a rotor blade is varied. Sometimes referred to as the spanwise axis.

FEEDBACK—The transmittal of forces, which are initiated by aerodynamic action on rotor blades, to the cockpit controls.

FLAPPING HINGE—The hinge that permits the rotor blade to flap and thus balance the lift generated by the advancing and retreating blades.

FLAPPING—The vertical movement of a blade about a flapping hinge.

FLARE—A maneuver accomplished prior to landing to slow down a rotorcraft.

FREE TURBINE—A turboshaft engine with no physical connection between the compressor and power output shaft.

FREEWHEELING UNIT—A component of the transmission or power train that automatically disconnects the main rotor from the engine when the engine stops or slows below the equivalent rotor r.p.m.

FULLY ARTICULATED ROTOR SYSTEM—See articulated rotor system.

GRAVITY—See weight.

GROSS WEIGHT—The sum of the basic empty weight and useful load.

GROUND EFFECT—A usually beneficial influence on rotorcraft performance that occurs while flying close to the ground. It results from a reduction in upwash, downwash, and bladetip vortices, which provide a corresponding decrease in induced drag.

GROUND RESONANCE—Self-excited vibration occurring whenever the frequency of oscillation of the blades about the lead-lag axis of an articulated rotor becomes the same as the natural frequency of the fuselage.

GYROCOPTER—Trademark applied to gyroplanes designed and produced by the Bensen Aircraft Company.

GYROSCOPIC PRECESSION—An inherent quality of rotating bodies, which causes an applied force to be manifested 90° in the direction of rotation from the point where the force is applied.

HUMAN FACTORS—The study of how people interact with their environment. In the case of general aviation, it is the study of how pilot performance is influenced by such issues as the design of cockpits, the function of the organs of the body, the effects of emotions, and the interaction and communication with other participants in the aviation community, such as other crew members and air traffic control personnel.

HUNTING—Movement of a blade with respect to the other blades in the plane of rotation, sometimes called leading or lagging.

INERTIA—The property of matter by which it will remain at rest or in a state of uniform motion in the same direction unless acted upon by some external force.

IN GROUND EFFECT (IGE) HOVER—Hovering close to the surface (usually less than one rotor diameter distance above the surface) under the influence of ground effect.

INDUCED DRAG—That part of the total drag that is created by the production of lift.

INDUCED FLOW—The component of air flowing vertically through the rotor system resulting from the production of lift.

ISOGONIC LINES—Lines on charts that connect points of equal magnetic variation.

KNOT—A unit of speed equal to one nautical mile per hour.

L/D$_{MAX}$—The maximum ratio between total lift (L) and total drag (D). This point provides the best glide speed. Any deviation from the best glide speed increases drag and reduces the distance you can glide.

LATERAL VIBRATION—A vibration in which the movement is in a lateral direction, such as imbalance of the main rotor.

LEAD AND LAG—The fore (lead) and aft (lag) movement of the rotor blade in the plane of rotation.

LICENSED EMPTY WEIGHT—Basic empty weight not including full engine oil, just undrainable oil.

LIFT—One of the four main forces acting on a rotorcraft. It acts perpendicular to the relative wind.

LOAD FACTOR—The ratio of a specified load to the total weight of the aircraft.

MARRIED NEEDLES—A term used when two hands of an instrument are superimposed over each other, as on the engine/rotor tachometer.

MAST—The component that supports the main rotor.

MAST BUMPING—Action of the rotor head striking the mast, occurring on underslung rotors only.

MINIMUM LEVEL FLIGHT SPEED—The speed below which a gyroplane, the propeller of which is producing maximum thrust, loses altitude.

NAVIGATIONAL AID (NAVAID)—Any visual or electronic device, airborne or on the surface, that provides point-to-point guidance information, or position data, to aircraft in flight.

NIGHT—The time between the end of evening civil twilight and the beginning of morning civil twilight, as published in the American Air Almanac.

NORMALLY ASPIRATED ENGINE—An engine that does not compensate for decreases in atmospheric pressure through turbocharging or other means.

ONE-TO-ONE VIBRATION—A low frequency vibration having one beat per revolution of the rotor. This vibration can be either lateral, vertical, or horizontal.

OUT OF GROUND EFFECT (OGE) HOVER—Hovering greater than one diameter distance above the surface. Because induced drag is greater while hovering out of ground effect, it takes more power to achieve a hover out of ground effect.

PARASITE DRAG—The part of total drag created by the form or shape of helicopter parts.

PAYLOAD—The term used for passengers, baggage, and cargo.

PENDULAR ACTION—The lateral or longitudinal oscillation of the fuselage due to it being suspended from the rotor system.

PITCH ANGLE—The angle between the chord line of the rotor blade and the reference plane of the main rotor hub or the rotor plane of rotation.

PREROTATION—In a gyroplane, it is the spinning of the rotor to a sufficient r.p.m. prior to flight.

PRESSURE ALTITUDE—The height above the standard pressure level of 29.92 in. Hg. It is obtained by setting 29.92 in the barometric pressure window and reading the altimeter.

PROFILE DRAG—Drag incurred from frictional or parasitic resistance of the blades passing through the air. It does not change significantly with the angle of attack of the airfoil section, but it increases moderately as airspeed increases.

RESULTANT RELATIVE WIND—Airflow from rotation that is modified by induced flow.

RETREATING BLADE—Any blade, located in a semicircular part of the rotor disc, where the blade direction is opposite to the direction of flight.

RETREATING BLADE STALL—A stall that begins at or near the tip of a blade in a helicopter because of the high angles of attack required to compensate for dissymmetry of lift. In a gyroplane the stall occurs at 20 to 40 percent outboard from the hub.

RIGID ROTOR—A rotor system permitting blades to feather but not flap or hunt.

ROTATIONAL VELOCITY—The component of relative wind produced by the rotation of the rotor blades.

ROTOR—A complete system of rotating airfoils creating lift for a helicopter or gyroplane.

ROTOR DISC AREA—See disk area.

ROTOR BRAKE—A device used to stop the rotor blades during shutdown.

ROTOR FORCE—The force produced by the rotor in a gyroplane. It is comprised of rotor lift and rotor drag.

SEMIRIGID ROTOR—A rotor system in which the blades are fixed to the hub but are free to flap and feather.

SETTLING WITH POWER—See vortex ring state.

SHAFT TURBINE—A turbine engine used to drive an output shaft commonly used in helicopters.

SKID—A flight condition in which the rate of turn is too great for the angle of bank.

SKID SHOES—Plates attached to the bottom of skid landing gear protecting the skid.

SLIP—A flight condition in which the rate of turn is too slow for the angle of bank.

SOLIDITY RATIO—The ratio of the total rotor blade area to total rotor disc area.

SPAN—The dimension of a rotor blade or airfoil from root to tip.

SPLIT NEEDLES—A term used to describe the position of the two needles on the engine/rotor tachometer when the two needles are not superimposed.

STANDARD ATMOSPHERE—A hypothetical atmosphere based on averages in which the surface temperature is 59°F (15°C), the surface pressure is 29.92 in. Hg (1013.2 Mb) at sea level, and the temperature lapse rate is approximately 3.5°F (2°C) per 1,000 feet.

STATIC STOP—A device used to limit the blade flap, or rotor flap, at low r.p.m. or when the rotor is stopped.

STEADY-STATE FLIGHT—A condition when a rotorcraft is in straight-and-level, unaccelerated flight, and all forces are in balance.

SYMMETRICAL AIRFOIL—An airfoil having the same shape on the top and bottom.

TAIL ROTOR—A rotor turning in a plane perpendicular to that of the main rotor and parallel to the longitudinal axis of the fuselage. It is used to control the torque of the main rotor and to provide movement about the yaw axis of the helicopter.

TEETERING HINGE—A hinge that permits the rotor blades of a semi-rigid rotor system to flap as a unit.

THRUST—The force developed by the rotor blades acting parallel to the relative wind and opposing the forces of drag and weight.

TIP-PATH PLANE—The imaginary circular plane outlined by the rotor blade tips as they make a cycle of rotation.

TORQUE—In helicopters with a single, main rotor system, the tendency of the helicopter to turn in the opposite direction of the main rotor rotation.

TRAILING EDGE—The rearmost edge of an airfoil.

TRANSLATING TENDENCY—The tendency of the single-rotor helicopter to move laterally during hovering flight. Also called tail rotor drift.

TRANSLATIONAL LIFT—The additional lift obtained when entering forward flight, due to the increased efficiency of the rotor system.

TRANSVERSE-FLOW EFFECT—A condition of increased drag and decreased lift in the aft portion of the rotor disc caused by the air having a greater induced velocity and angle in the aft portion of the disc.

TRUE ALTITUDE—The actual height of an object above mean sea level.

TURBOSHAFT ENGINE—A turbine engine transmitting power through a shaft as would be found in a turbine helicopter.

TWIST GRIP—The power control on the end of the collective control.

UNDERSLUNG—A rotor hub that rotates below the top of the mast, as on semirigid rotor systems.

UNLOADED ROTOR—The state of a rotor when rotor force has been removed, or when the rotor is operating under a low or negative G condition.

USEFUL LOAD—The difference between the gross weight and the basic empty weight. It includes the flight crew, usable fuel, drainable oil, if applicable, and payload.

VARIATION—The angular difference between true north and magnetic north; indicated on charts by isogonic lines.

VERTICAL VIBRATION—A vibration in which the movement is up and down, or vertical, as in an out-of-track condition.

VORTEX RING STATE—A transient condition of downward flight (descending through air after just previously being accelerated downward by the rotor) during which an appreciable portion of the main rotor system is being forced to operate at angles of attack above maximum. Blade stall starts near the hub and progresses outward as the rate of descent increases.

WEIGHT—One of the four main forces acting on a rotorcraft. Equivalent to the actual weight of the rotorcraft. It acts downward toward the center of the earth.

YAW—The movement of a rotorcraft about its vertical axis.

INDEX

N

O

P

Q

R

S

T